Sustainable Development Goals Series

The **Sustainable Development Goals Series** is Springer Nature's inaugural cross-imprint book series that addresses and supports the United Nations' seventeen Sustainable Development Goals. The series fosters comprehensive research focused on these global targets and endeavours to address some of society's greatest grand challenges. The SDGs are inherently multidisciplinary, and they bring people working across different fields together and working towards a common goal. In this spirit, the Sustainable Development Goals series is the first at Springer Nature to publish books under both the Springer and Palgrave Macmillan imprints, bringing the strengths of our imprints together.

The Sustainable Development Goals Series is organized into eighteen subseries: one subseries based around each of the seventeen respective Sustainable Development Goals, and an eighteenth subseries, "Connecting the Goals," which serves as a home for volumes addressing multiple goals or studying the SDGs as a whole. Each subseries is guided by an expert Subseries Advisor with years or decades of experience studying and addressing core components of their respective Goal.

The SDG Series has a remit as broad as the SDGs themselves, and contributions are welcome from scientists, academics, policymakers, and researchers working in fields related to any of the seventeen goals. If you are interested in contributing a monograph or curated volume to the series, please contact the Publishers: Zachary Romano [Springer; zachary.romano@springer.com] and Rachael Ballard [Palgrave Macmillan; rachael.ballard@palgrave.com].

More information about this series at http://link.springer.com/series/15486

Sandeep Tambe

Teaching and Learning Rural Livelihoods

A Guide for Educators, Students, and Practitioners

Springer

Sandeep Tambe
Indian Institute of Forest Management
Bhopal, Madhya Pradesh, India

ISSN 2523-3084 ISSN 2523-3092 (electronic)
Sustainable Development Goals Series
ISBN 978-3-030-90490-6 ISBN 978-3-030-90491-3 (eBook)
https://doi.org/10.1007/978-3-030-90491-3

This Springer imprint is published by the registered company Springer Nature Switzerland AG
The registered company address is: Gewerbestrasse 11, 6330 Cham, Switzerland

Inspired by the life of Mahatma Gandhi
who in his own gentle way shook
the world.

Preface

This book aspires to strengthen the livelihoods of millions of people left behind in our quest for growth and globalization. It aims to do this by synthesizing the existing knowledge on livelihoods to better inform the science, policy, and practice of ending poverty and hunger. In doing so, we broaden the scope of the term *rural livelihoods* to include access to assets, infrastructure, goods and services, reducing vulnerability, strengthening governance, breaking down structural barriers, enabling rights, fulfilling obligations, promoting equitable access to opportunities and freedoms, and safeguarding from shocks and stresses. Over the last five decades, rural livelihoods as a discipline has received a seminal contribution from some of the greatest minds of our times, whose scholarly work has helped conceptualize the rural livelihood approaches. These approaches form the central theme of this book, and we explore the contribution they can make in assessing, planning, strategizing, monitoring, and evaluating livelihood studies, livelihood policy, and livelihood programmes. While these approaches hold immense potential in enriching our understanding of livelihoods, ironically they have received less attention from policymakers, donors, and development professionals. The purpose of this book is to assimilate these approaches, analyse them, demonstrate their application, showcase their value, synthesize them, and unify them as a family of approaches.

In this book, we track the evolution of rural livelihoods thinking over the last five decades and find that rural livelihoods as a discipline has received a pioneering contribution from global icons and devoted institutions. They helped define, redefine, and broaden the boundaries of this field while also deepening the conceptual part. These contributions played a pivotal role in advancing our understanding of livelihoods and poverty. When we discuss rural livelihoods, typically the DFID sustainable livelihoods approach intuitively comes to mind. In this book, the term *rural livelihood approach* is used in a broader sense and encompasses several other analytical approaches on rural livelihoods that have been conceptualized, applied, and discussed over the last five decades. These include the sustainable livelihoods approach, common pool resources, livelihood trajectories, rights-based approach, graduation approach, and the resilience framework. We discuss these six livelihood approaches in detail and opine that they need to be imagined as a family of livelihood approaches and depending upon the context used either in solitary mode or in combination.

The global development agenda over the last few decades has been dominated by the Millennium Development Goals (MDGs) and the Sustainable Development Goals (SDGs). The period 1990–2015 saw a rapid decline in global poverty, and the MDGs on halving extreme poverty were met ahead of schedule. This change was driven by structural adjustments in the economy, market liberalization, and rural transformation, supplemented by a redistributive policy. The developing nations in Asia and Africa witnessed a transition from subsistence level farming to livelihood diversification aided by rapid growth in non-farm employment, diverse livestock ownership, and migration. This lead to a structural change in the economy of these countries from agriculture to industry and services. However, the progress was unequal with several countries in Sub-Saharan Africa and South Asia reporting high poverty levels and weak progress in non-income goals related to health, education, and living standards. This period also saw widespread devastation of nature and disruption of the earth's support systems putting a question mark on the very survival of humankind. A course correction came in the form of the SDGs, which focussed both on the well-being of people and our planet. The SDGs adopted an ambitious target of eliminating extreme poverty and hunger in all forms everywhere by 2030. They thus put the developmental spotlight back on poverty and livelihoods with an added element of environmental sustainability, combating entrenched poverty, and also building resilience.

In courses on developmental studies, the rural livelihood approaches are often drawn upon extensively to enliven and enrich the class discussion. The 'missing link' is an academic synthesis that assimilates these seminal works under one cover. Conceptualizing a rural livelihoods book that focusses on theory and practice both has several challenges. The teaching material has to be structured to cover the conceptual as well as the applied part. Class exercises have to be included to simulate the application of theory in practice. Real-life scenarios have to be recreated to demonstrate the integration of the learnings in practice. Presently, there is a dearth of rural livelihood textbooks to refer to, and educators have to devise their own ways to address these issues while teaching this subject. Course offerings on rural livelihoods are also yet to figure in the popular online learning platforms. Hence, development students are left to sift through a myriad of articles to inform their practice. Over the past few years, the attention to rural livelihoods is waning due to the broad processes of global change that seem to be redirecting attention to urban studies and macro-structural issues.

A textbook that synthesizes the conceptual, analytical, and applied aspects in one place will lighten the task of the teacher and student alike. It will make the teaching and learning of this subject more scientific, systematic, and applied. Bridging the gap between theory and practice will also make the students field-ready and enable them to add value to the institutions they join right away. While there is a sizable audience that stands to benefit, 'rural livelihood' is presently an under-published area. This textbook aspires to occupy this vacant niche and make a valuable contribution to this field of knowledge. The purpose behind conceptualizing this textbook is to provide a strong foundation for educators and practitioners alike. It has been conceived and produced for the broader development community. It synthesizes the conceptual aspects of rural livelihood approaches and their practice in development. The book draws as much from practice as it draws from theory. It focusses both on the *what* and the *how-to* part, that is *what* are the academic advancements in livelihood thinking and *how to* apply these concepts in practice?

What this book does not cover though is the thematic areas of education, health, public infrastructure, and the like which like livelihoods are equally vital for human well-being. Though these aspects are critical for sustainable rural development, they are beyond the scope of this book. The objective of this book is to explore pathways for strengthening rural livelihoods that can fructify in the short- and mid-term, typically within a span of 3–10 years. In terms of scale, the book has a wider range and the learnings have the potential to inform the theory of change of local projects, aid the design of livelihood interventions at the sub-national level, and also help conceptualize livelihood policies at the national level. Hence, while the importance and inter-relatedness of the other SDGs on education, health, living standards, economic growth, secure jobs, environment, etc. are discussed and drawn upon, the positioning of this book strongly aligns with the SDG 1 of ending poverty in all its forms everywhere.

The book is a result of a collective journey with development professionals, academicians, policymakers, and students. It draws on years of developmental practice and experience gained in teaching rural livelihoods to graduate students. It hopes to develop a connection between the developments in livelihoods thinking, livelihood approaches, and developmental practice. At present, rural development projects are dominated by the usage of general planning approaches and rural livelihood approaches are not adequately informing practice. The objective of this book is to bring the concepts, frameworks, models, and theories related to rural livelihoods under one cover. The textbook is positioned as a one-stop destination for those engaged in the science, policy, and practice of rural livelihoods. It attempts to synthesize the scholarship on rural livelihoods in one place and guide students and young professionals to start using these analytical approaches in their research and development work. Several exercises and box items are included in the chapters to make the book user-friendly. Practitioners prefer clear messages without much ambiguity or complexity. They have a reductionist approach to summarize and simplify information so that it is comprehensible to policymakers, front-line staff, and the local communities. Readers can discern that this book is from this genre and written by a practitioner.

Bhopal, Madhya Pradesh, India Sandeep Tambe

Acknowledgements

The book builds on, assimilates, and synthesizes the seminal work of several eminent academicians and practitioners who have developed concepts, frameworks, and models to unravel the complexities of rural livelihoods. It also draws heavily from several years of practice during my stint in the Department of Rural Development, in the Himalayan state of Sikkim, India. Here I gained hands-on experience while managing a diverse portfolio of rural development programmes comprising socio-economic surveys, public works, rural housing, rural infrastructure, water security, sanitation, post-disaster reconstruction, decentralization, rural planning, social audits, capacity building, extension, concurrent monitoring, and external evaluation. The book would not have been possible without the contribution of a team of civil service officials, elected representatives, technical experts, consultants, and funding agencies who shared this journey of occasional lows, excruciating reverses, and noble triumphs. The team struck it together and developed a unique bond and a shared vision creating several innovations and novelties that would later be recognized and adopted for national scale-up.

The academic part however happened after the practice, when after my field experience I took up tenure as a faculty at the Indian Institute of Forest Management. Here, free from the pressures of programme delivery, I got the opportunity to delve into academics. Hence, the theorizing part came after the application, when I was able to explore the present state of knowledge in poverty alleviation and livelihoods from the standpoint of how it could better inform practice. An elective that I offered and was much subscribed to by the graduate students of management was on Applied Rural Planning and Livelihoods and helped to frame the science, policy, and practice of livelihoods into a teaching and learning module. The experience of developing this course and teaching for several years helped me to also realize the dearth of a readymade textbook for this subject and the trials and tribulations that educators and students have to go through in collating the teaching resources that lie scattered in books, reports, journal papers, and websites of development organizations. It was only while offering this elective for the sixth consecutive year that I decided to sum up my learnings as a practitioner and then as an educator in a book form. So in that sense, the book draws heavily from my diverse stints, first as a rural development professional and subsequently as a faculty.

The conversation about this book started towards the end of March 2020, when Aaron Shiller the Assistant Editor with Springer Nature contacted me to review a book proposal on the Himalayas. The reviewer questionnaire towards the end had a question, 'Would you be interested in writing or editing a volume now or in the future'? This question set me thinking. Novel Coronavirus was spreading rapidly around the globe and India was in the midst of a complete lockdown. My field research projects had all come to a halt. I thought maybe this was a godsend opportunity to step back, assimilate, and synthesize a rural livelihoods book. A phone call from Aaron helped clarify the modalities of book publishing. The academic dean at my institute—Professor C. S. Rathore and the Director Pankaj Srivastava responded enthusiastically to this idea. By August 2020, I had submitted the book proposal to Springer, and by early October we received encouraging comments from the anonymous reviewers. The book contract was signed in early November 2020 and the first draft of the book was completed and submitted in August 2021.

I owe loads of gratitude to the authors, institutions, and publishers who generously permitted us to share excerpts, tables, and figures from their publications to enrich this book. I am grateful to K. V. C. Sekhar for taking out time and lending his expertise to design and digitize the illustrations.

This book owes its existence to the contribution of my teachers, supervisors, family, colleagues, and students. They encouraged and supported this initiative in multifarious ways. Their queries, constant encouragement, and sharp feedback helped to spark and fuel this initiative. I am indebted to Himanshu Rohira who edited all the chapters and also improved their readability. A continuous engagement with him spanning across several months helped in smoothening the rough edges and improving the overall quality of the manuscript. In any journey, it is your loved ones, whose eternal trust and confidence keep you going through the highs and lows. My parents G.N. Tambe and Sheela Tambe, and my sisters Sanjana Patrikar and Deepali Pangaonkar have been always very supportive of my travails often on unconventional paths. My grandparents, N.K. Tambe and Indirabai Tambe, uncle A.N. Tambe, aunt Kusum Tambe, and aunt Vijaya Pandit hosted my summer vacations in a remote forest fringe village in central India. As an urban kid, this lived experience in an agrarian hamlet with indigenous communities, rolling hills, tranquil rivers, dense forests, ancient forts, and vibrant farm life developed into a lifelong fascination for the 'rural'. My wife Trupti, son Dhruv, and pet dog Nimbus provided a loving nest, and the pristine IIFM campus an ideal habitat, where this idea could incubate, find its wings, and take flight. Their love and enduring support have been central in this book taking shape and getting completed.

Bhopal, Madhya Pradesh, India Sandeep Tambe

About the Book

The textbook is divided into three parts comprising of 13 chapters as follows:

Part I: Introduction to Rural Livelihoods (Chaps. 1–3)

Chapter 1 traces the roots and evolution of rural livelihoods as a theme led by seminal contributions from some of the finest minds and committed institutions of our times. We discuss the genesis, growth, and blossoming of rural livelihoods over the last five decades as an academic discipline and its contribution to development thinking. A synopsis of the six rural livelihood approaches is provided along with their key elements. Surprisingly in spite of these advancements in our understanding of rural livelihoods and poverty, developmental initiatives often take recourse to generalistic project management approaches. We opine that there is a need for a stronger connect and uptake of the science of livelihoods both in policy and practice. We urge policymakers and development practitioners to build on these livelihood approaches while constructing the logframe matrix, theory of change, conceptualizing rural livelihood programmes, and poverty policies.

In Chap. 2, we discuss the rapid structural change sweeping across the global south. In the past in rural areas, food crop agriculture and livestock husbandry were the mainstay of the rural economy. The process of structural adjustment and market liberalization unleashed a powerful process of change affecting economies and societies worldwide. These changes were characterized by rising agricultural productivity, change in the composition of the economy from a dependence on agriculture to manufacturing and services, integration with global markets, rural to urban migration, and rapid urbanization. Consequently, the share of the agriculture sector to the economy declined rapidly. The better-off rural households successfully attained sustainable asset accumulation by taking up non-farm activities to supplement their farm-based incomes. However, in many countries in South Asia and Sub Saharan Africa (SSA), this structural transformation is incomplete due to weak industrialization, and there is a persistence of smallholder farming with agriculture still contributing sizeably to employment. As highlighted by the SDGs, economic growth that creates full employment and decent work with equal pay for work of equal value is at the heart of inclusive and poverty reduction growth. While structural adjustments and liberalization are necessary for inclusive rural transformation, these alone are not sufficient. We opine that these changes need to be supplemented with redistributive policy and programmatic measures to enable the rural poor to seize opportunities while adapting to the risks and changes.

In Chap. 3, we discuss the continued relevance of rural livelihoods as a discipline and place it in the context of the global development agenda. We analyse the significance of rural livelihoods in meeting the SDG goals of ending poverty and hunger in all forms everywhere. During the MDG era, significant achievements were made in halving global poverty, powered mainly by large Asian countries. The SDG era of poverty alleviation will be much more challenging as the action shifts to smaller

African countries with complex structural issues. We opine that while significant progress has been made in halving global poverty in this century, extreme poverty still persists and there is a need to maintain the momentum to eradicate global poverty sustainably. We argue that even in the post-2030 world, growing risks and vulnerabilities will test the sustainability of the SDG attainments, keeping rural livelihoods relevant in the future as well. However, over the last decade, the interest in rural livelihoods is waning and it appears to be entering a senescence phase. We urge the development community to renew their interest in the growth and evolution of rural livelihoods to address future challenges in poverty alleviation.

Part II: Synthesis of the Present State of Knowledge (Chapters 4–10)

This part forms the core of this book where we describe the analytical frameworks of the various rural livelihood approaches. We assimilate the six main rural livelihood approaches of the last five decades. We describe the sustainable livelihoods approach, common pool resources, livelihood trajectories, rights-based approach, graduation approach, and resilience framework. For each of these analytical frameworks, we discuss the conceptual part, its constituent elements, the process it prescribes, its application in practice, and the insights it unravels. This part is specially designed keeping in mind the requirements of educators and students in teaching and learning rural livelihoods. The teachers need to evaluate student learning, while the students need to practice and prepare for the assessments. The exercises section has been provided in all the chapters and aims to sharpen the ability of the students to apply these approaches and interpret the findings to inform policy and practice. These class exercises consist of research articles, problems, and quizzes that test the analytical ability of the students. The students can attempt these exercises to further hone their skills in applying the livelihood approaches. The last chapter provides a comparative analysis where we assess these livelihood approaches against a set of objective parameters. The aim is to develop a deeper understanding of their comparative strengths and weaknesses, to be able to analyse which approach to use when and what value it can add.

Part III: Strengthening the Knowledge Action Pathway (Chapters 11–13)

This part engages with the situations, issues, and challenges practitioners face and how the science of rural livelihoods can better inform practice. What role livelihood approaches can play in guiding the planning, execution, and evaluation of rural development projects? How they can inform, guide, and add value to the design of a livelihood development proposal? Can they inform what is working and what is not in a rural setting? In this part we provide guidance on how to apply and adapt the various rural livelihood approaches in real-life situations. We also discuss the various challenges that may emerge in implementation and how to address them. The practice of rural development is still dominated by generalist approaches such as the logical framework approach, theory of change, and various monitoring and evaluation frameworks. We explore how to integrate these rural livelihood approaches with these generalist planning approaches to mainstream them in development. Finally, in the way forward we discuss the new elements and contexts the livelihoods concept needs to assimilate to stay relevant and guide development in the coming decades.

The purpose of this book is to evolve a shared understanding of rural livelihoods and help connect the different worlds of academia, development professionals, and policymakers. The objective of this work is to assimilate the present state of knowledge on rural livelihoods, collate it as a bouquet of distinct approaches, ground these approaches in practice with the help of exercises and examples, and creatively reflect on them to accelerate their uptake in science, policy, and practice. The expected outputs from penning this textbook are threefold. Firstly, it positions itself as a one-stop destination for educators by assimilating the rural livelihood works of eminent academicians under one cover. Secondly, it provides a step-by-step approach to apply these learnings in real-life situations. Thirdly, it translates these concepts into a ready-to-teach applied format by generously adding class exercises from practice and thus functions as a ready reckoner for professionals engaged in delivering rural livelihoods. The expected outcomes from penning this textbook are also threefold. One to provide a solid foundation to the teaching and learning of rural livelihoods in academia by offering a syllabus which the educator can build upon. Secondly, to build a bridge between the science, policy, and practice of rural livelihoods. And finally, to shape a pool of better informed and equipped professionals in development.

In terms of its potential audience, the book has a wide range. It will appeal to teachers, students, development practitioners, government departments, policymakers, public service academies, training institutes, funding agencies, charitable trusts, philanthropic foundations, NGOs, civil society organizations, international development agencies, and others. Various academic courses covering the themes of international development, development studies, environment and development, sustainable development, poverty and inequity, power relations, rural livelihoods, etc. will find this book especially useful. If this book can create a better understanding of the theory and practice of rural livelihoods and shape a pool of better informed and skilled professionals, then its purpose would have been met.

Contents

About the Author

Sandeep Tambe is a member of the Indian Forest Service. He holds a doctoral degree in forest management from the Wildlife Institute of India and a graduate degree in engineering from the Indian Institute of Technology, Mumbai. He has multi-disciplinary expertise in physical, natural, and social sciences and management disciplines. Having served in government, civil society, and academic setups has enabled him to better understand the worlds of science, policy, and practice. He is passionate about nature conservation, strengthening rural livelihoods, science-policy-practice interface, and democratic governance of the commons. He served in the Rural Development Department in the Government of Sikkim from 2007 to 2015, where he gained hands-on experience in managing diverse rural development programmes. He embedded a people-centric, results-based, and scientific approach in these programmes, and many of these innovations were recognized and endorsed for national mainstreaming. Following his stint in practice and shaping policy, he joined as a professor at the Indian Institute of Forest Management (IIFM) in 2015. Here he got the opportunity to study rural livelihoods as an academic discipline and also teach it to graduate students. He has published 21 research papers and 5 books. He was selected as the top 25 persons of the year by the Forbes India magazine in 2010. He was conferred the T.N. Khoshoo Memorial Award in 2012 for his efforts in sustainability and community-based governance of natural resources and the State Award for Meritorious Service in 2014. The livelihood programmes managed by him won the Prime Minister's Award for Excellence in Public Administration in 2013, National Award for Excellence in Convergence in 2015, and the National Award for Transparency and Accountability in 2015. In 2017, the IIFM student council selected him for the Best Teacher Award.

Abbreviations

AIDS	Acquired Immune Deficiency Syndrome
ASP	Adaptive social protection
BRAC	Bangladesh Rural Advancement Committee
CBNRM	Community-based natural resource management
CBO	Community-based organization
CCA	Climate change adaptation
CFR	Community forest resource rights under FRA
CGAP	Consultative Group to Assist the Poor
CIF	Community investment fund
CPR	Common pool resources
DFID	Department for International Development, Government of United Kingdom
DRR	Disaster risk reduction
FAO	Food and Agriculture Organization
FGD	Focus group discussion
FRA	Scheduled Tribes and Other Traditional Forest Dwellers (Recognition of Forest Rights) Act, 2006
GA	Graduation approach
GDP	Gross domestic product
HLAs	Household livelihood assets
IDS	Institute of Development Studies
IFAD	International Fund for Agricultural Development
IGA	Income generation activities
IIFM	Indian Institute of Forest Management
IPBES	The Intergovernmental Science-Policy Platform on Biodiversity and Ecosystem Services
KII	Key informant interview
KPI	Key performance indicators
LFA	Logical framework approach
MDG	Millennium development goals
MFI	Microfinance institution
MGNREGA	Mahatma Gandhi National Rural Employment Guarantee Act
MORD	Ministry of Rural Development, Government of India
MPI	Multidimensional poverty index
NGO	Non-government organization
NRLM	National Rural Livelihood Mission, Government of India
NRM	Natural resource management
NTFP	Non-timber forest product
ODI	Overseas Development Institute

OECD	Organisation for Economic Co-operation and Development
OPHI	Oxford Poverty and Human Development Initiative
PCM	Project cycle management
PIP	Policies, institutions, and processes
PLA	Participatory learning and action
PRA	Participatory rural appraisal
PWR	Participatory well-being ranking
RBA	Rights-based approach
RBM	Results-based management
RF	Revolving fund
RLA	Rural livelihood approaches
RDD	Rural Development Department, Government of Sikkim
RNF	Rural non-farm
RRA	Rapid rural appraisal
SDG	Sustainable development goals
SES	Socio-ecological systems
SHG	Self-help group
SLA	Sustainable livelihoods approach
SLF	Sustainable livelihoods framework
SOP	Stages of progress
SPPI	Science, policy, and practice interface
SRL	Sustainable rural livelihood
SSA	Sub-Saharan Africa
SWOT	Strength, weakness, opportunity, and threat
TB	Tuberculosis
TOC	Theory of change
UN	United Nations
UNDESA	United Nations Department of Economic and Social Affairs
UNDP	United Nations Development Programme
UNDP HDR	UNDP Human Development Report

List of Figures

List of Tables

List of Boxes

Part I
Introduction to Rural Livelihoods

The first section of this book provides an overview of rural livelihoods as a discipline and its continued relevance in the context of global environmental change, structural reforms, and sustainable development. We begin by tracing the trends in rural development thinking, the evolution of the various rural livelihood approaches, their key elements, and their significance. We then take a macro look at structural reforms and rural transformation and how these forces have impacted small holder farming, non-farm employment, migration, and poverty reduction. Towards the end, we trace the global commitment to reduce poverty reflected in the millennium development goal partnership, the achievements during the MDG period, and the need for the SDGs to tackle the unfinished agenda of uneven progress, inequity, and environmental degradation. We opine that a continued emphasis on rural livelihoods is needed for ending poverty during the SDG period and for sustaining it thereafter.

Chapter 1
Evolution of Rural Livelihood Approaches

Abstract This chapter discusses the genesis, growth, and prospering of the rural livelihoods discipline over the last five decades. We start by describing the evolution of rural development thinking that lend the context to rural livelihoods. We then analyse the evolution of rural livelihood approaches and the notable contributions made by eminent academicians, practitioners, and institutions in unravelling its various dimensions. This is followed by an introduction to the six key livelihood approaches, namely sustainable livelihoods approach, common pool resources, livelihood trajectories, rights-based approach to livelihoods, graduation approach for the ultra-poor, and the resilience framework. We also explore the trends in development studies and the inter-linkages with livelihood approaches. We opine that despite the academic advancements in rural livelihoods, it has not been able to sufficiently influence science, policy, and practice. We urge researchers, policymakers, and the development community to actively participate in the growth of the rural livelihoods domain, its integration with new emerging areas, and its adoption in practice.

Keywords Livelihood perspectives · Development thinking · Livelihood approaches · Evolution · Sustainable · Dynamics · Commons · Graduation · Resilience · Development studies

1.1 Trends in Rural Development Thinking

Over the years, the discourse on development has been evolving and diverse ideas have been debated, applied, persisted with, and also often discarded. The understanding of rural development as a concept and its various dimensions has transformed over the last five decades. It has evolved from a structural perspective which looked at the state playing the role of a benefactor, to an actor-based perspective that looks at the poor as proactive agents of transformation, to a rights-based approach with normative concerns to sustainability issues (de Haan & Zoomers, 2005; Bohle, 2009). We discuss these perspectives in greater detail below (Fig. 1.1).

Before the 1980s, the neo-Marxist perspective was prevalent, where rural households were perceived as lacking agency and incapable of securing their livelihoods without external intervention by the state using sectoral approaches (de Haan & Zoomers, 2005). In this the post-colonial era, development thinking was based on the virtues of the 'welfare state' (Box 1.1) putting the onus on the state as a benefactor to drive the development agenda with interventionist policies while constructing people as passive recipients in a largely protectionist world (Opschoor et al., 2005). During this era the development strategy in rural areas focussed on sectoral approaches such as modernizing agriculture, irrigation development, green revolution, farm mechanization, etc., which were planned in a top-down, technical manner that treated the participants as hapless recipients of aid. It was assumed that

S. Tambe, *Teaching and Learning Rural Livelihoods*, Sustainable Development Goals Series,
https://doi.org/10.1007/978-3-030-90491-3_1

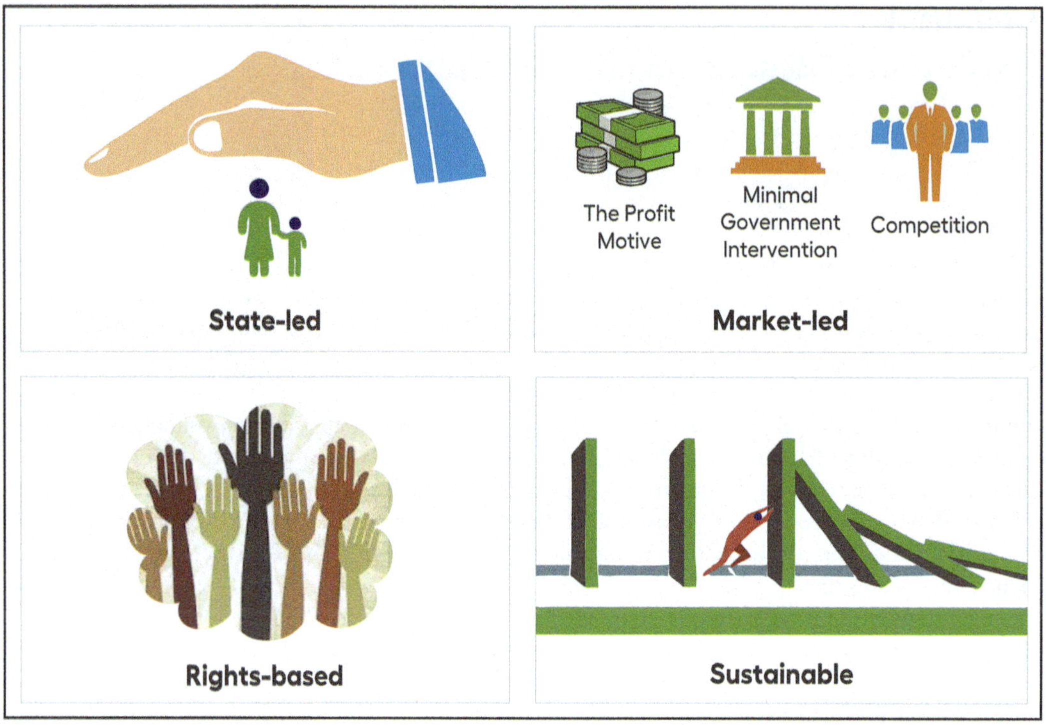

Fig. 1.1 Trends in rural development thinking over the last five decades

these macro-economic policies would uniformly improve the quality of life (Bagchi et al., 1998). The rural development thinking was based on the three paradigms of production, employment, and income poverty (Bohle, 2009). The developmental challenges of hunger, malnutrition, and famine were directly linked with inadequate food production but were later found to be attributed to problems with access, benefit sharing, and rights, rather than famines and supply shortages (Sen, 1981). The employment problem assumed that the rural people are jobless or unemployed, while they were in fact actively seeking livelihood opportunities and adapting to them (Bohle, 2009). Also, the poverty that rural people perceive and experience is not only in monetary terms, but also in the form of other deprivations in health, education, and living standards (Alkire & Santos, 2014).

Box 1.1 What Is a 'Welfare State'?

A welfare state is an expression used by political economy scholars to refer to the phenomenon of state-led macro-economic interventions that started in the late twentieth century, in the post-colonial period. In this model of capitalism, the state has more independent and autonomous political power, as well as dominance over the economy. A welfare or development state is characterized by having strong state intervention along with extensive regulation and planning. The welfare state is often distinguished as a state that has a focus on economic development and one that takes necessary policy measures to accomplish its objective. A regulatory state on the other hand, in contrast to a welfare state, governs the economy mainly through regulatory agencies that are empowered to put in force, a variety of standards of behaviour to protect the public against market failures. A welfare state participates more directly in the economy through a

(continued)

Box 1.1 (continued)

variety of strategies to promote the growth of new industries and to facilitate the transition of investments from old to new and emerging industries. Governments in welfare states typically invest and mobilize capital into the emerging sector with an anticipation of having a maximum spillover effect for society at large. Despite all the evidence of the importance of a welfare state, the results of this model have been mixed. However, international aid agencies have stressed that without some degree of welfare state interventions and good governance, the trickle-down model of economic development may not be sufficient to eradicate poverty in the developing world.

By the late 1970s, it was realized that there was a considerable gap between the intent and impact of this state-led development. The limited success of this approach gave way to bottom-up, household-centric interventions such as watershed development, livelihood development, and the like, which were context specific and recognized the agency of the poor. This lead to the emergence in the 1980s of a neoliberal development agenda, banking on the virtues of a free-market-led development. It brought with it the actor-oriented perspective that valued the lived experience of the poor focussing on the micro-world of the household, its assets, its vulnerabilities, and its networks (Bohle, 2009). The actor-based perspective looks at the assets, capabilities, and opportunities the poor have, instead of what they do not have, and the proactive role they can play in exploiting these to secure their livelihoods (Bohle, 2009). This period also saw a growth of NGOs and adoption of participatory approaches in development. Structural adjustment coupled with deregulation and globalization became fashionable with accelerated economic growth being revered as the new messiah that would eradicate poverty. The 'trickle-down theory' gained currency with the belief that sustained growth would result in poverty reduction. However, towards the end of the twentieth century, there was a growing realization that this period of economic adjustment had resulted in income inequality which was the main factor behind persistent poverty. The global leaders adopted the broad-based Millennium Development Goals (MDGs) with a focus on human development and providing a new impetus to reducing poverty, hunger, and disease. It is now largely accepted that while globalization provided new opportunities, it did not benefit all sections of society, and a need was felt for more inclusive approaches.

With both the government-led and market-led approaches achieving modest success in addressing the deep-rooted structural inequalities that existed in society, a third perspective, namely the rights-based approach to livelihoods security started gaining prominence. With a shrinking welfare state and a realization that market friendly policies were leading to income disparities, it was felt that poverty alleviation was best approached from a rights-based perspective (Deka, 2012). This approach aims to address structural barriers that impede equitable growth and is based on the normative principles of participation, inclusion, equality, and accountability. It attempts to transform the patron client relationship between the state and its citizens in the traditional development assistance model to one where development is not looked upon as charity, but as a responsibility of the state as a duty bearer to fulfil the rights of its citizens (Broberg & Sano, 2018).

With rising impacts on the environment and human impacts reaching a planetary scale, there is a growing shift towards sustainability, which is further reinforced with the adoption of the Agenda 2030 for Sustainable Development. Climate change adaptation, disaster risk reduction, and building livelihood resilience in the face of rising crisis contexts are being prioritized. In the face of looming threats to human wellbeing from natural disasters, climate change, pandemics, economic crisis, and political upheavals, sustainability thinking has gained prominence in development debates. The global challenge is to achieve a sustainable future for all by 2030 with an emphasis to integrate environment with development (Elder & Olsen, 2019).

1.2 Conceptualizing Rural Livelihoods

Before getting into the discussion on rural livelihoods, it would be pertinent to acknowledge the phenomenal contribution of Professor Robert Chambers in the development of participatory approaches to learning and action that are now considered as the cornerstone of the development sector (Chambers, 1997, 2008). He is widely regarded as the father of rural development and for popularizing participatory approaches, bottom-up thinking, and local solutions (Box 1.2). He advocated for putting the poor and the local communities at the centre of the development process and the need for development professionals to work *with* and not *for* local communities. As a starting point, how would one define the term livelihood? Wikipedia defines *livelihood* as a '*means of securing the basic necessities (food, water, shelter, and clothing) of life*'. In 1987 the Brundtland Commission on Environment and Development pioneered the sustainable livelihoods concept. In 1992, this idea was further elaborated in the United Nations Conference on Environment and Development and advocated as the broad goal for poverty alleviation. In the same year, a composite definition of a sustainable rural livelihood was developed by Robert Chambers and Gordon Conway by building on the three concepts of capability, equity, and sustainability: '*A livelihood comprises of capabilities, assets (including both material and social resources), and activities required for a means of living. A livelihood is sustainable when it can cope with and recover from stresses and shocks and maintain or enhance its capabilities and assets both now and in the future, while not undermining the natural resource base*' (Chambers & Conway, 1992; Ashley & Carney, 1999; DFID, 1999). The theoretical underpinnings of the above definition can be traced to the Nobel laureate Amartya Sen's seminal work on capabilities and wellbeing. He describes *functionings* as the various activities that an individual takes up from amongst a larger set of *capabilities* one possesses and the *freedom* to choose a life one would like to lead (Sen, 1982, 1985, 1993). This freedom is however not absolute and is dependent on social arrangements and the larger policy environment provisioned by the state.

Box 1.2 Professor Robert Chambers—The Father of Rural Development

Professor Robert Chambers is one of the most significant scholars in international development studies. He has been a researcher at the Institute of Development Studies, the University of Sussex for the last four decades. Since the 1970s, he has published several books on the management of land settlement schemes and on rural development. His body of work drew on his earlier decade as an administrator, lecturer, and researcher in Africa and about 6 years of research in South Asia. He has been one of the leading voices for putting the poor, destitute, and marginalized at the centre of the processes of development. He is widely regarded as 'development's best advocate'. He advocated for the poor to be taken into account when the development problem is identified, a policy formulated and projects implemented. He popularized within development circles phrases such as 'putting the last first', 'survey slavery', 'simple is optimal', 'handing over the stick', etc. and stressed the now generally accepted need for development professionals to be critically self-aware. His techniques of participatory rural appraisal are widely accepted as a cornerstone of the rural development process to this day. Surprisingly, in the 1970s he was ignominiously turned down for a professorship at the Institute of Development Studies on the grounds of insufficient publications in scientific journals, despite his international fame. It was eventually in 1995 at the age of 63, not long before retirement, that he was awarded the professorship. His contributions have been recognized by multiple honorary doctorates awarded to him. In the 1960s, while teaching at the University of Nairobi in Kenya, Professor Chambers discovered his love for rock climbing. Throughout his career, he did multiple ascents especially in Kenya and later on in India.

1.3 Rural Livelihood Approaches

What is a livelihood approach? A livelihood approach is an analytical framework that aims to analyse and address poverty in all its dimensions. It enables the assessment of livelihood status, its dynamics, the associated opportunities and constraints, and in designing livelihood interventions to tackle poverty. It is applicable in diverse settings but more so in the global south. These approaches help us analyse the complexities of livelihoods—how people live in different settings, how they combine a web of activities to make a living, how these activities affect their livelihood outcomes, the longitudinal livelihood journeys they undertake, and the sustainability and resilience of these pathways (Scoones, 2009). Typically, in literature, the term livelihood approach is used synonymously with the DFID sustainable livelihood approach. In this book, we use this term more generically to collectively address a diverse set of livelihood approaches that have emerged over the last fifty years. During this period, rural livelihoods as a discipline received a seminal contribution from global icons such as Amartya Sen, Robert Chambers, Gordon R. Conway, Ian Scoones, Elinor Ostrom, Diana Carney, Leo de Haan, Fazle Hasan Abed, and several others. They have helped define, redefine, and broaden the boundaries of this field while also deepening the conceptual part. Several institutions such as IDS, DFID, IFAD, ODI, FAO, UNDP, Oxfam, CARE, World Bank, BRAC, and others have also played a stellar role in advancing our understanding of poverty and livelihoods. In this book, we focus on the six prominent approaches that help unearth and explain different facets of rural livelihoods and together help unpack its complexities (Fig. 1.2). We cover the sustainable livelihoods approach, collective governance of the commons, livelihood trajectories approach, rights-based approach, graduation approach, and the resilience approach. In this chapter, we discuss their salient features and delve deeper into these approaches in Part 2 of this book.

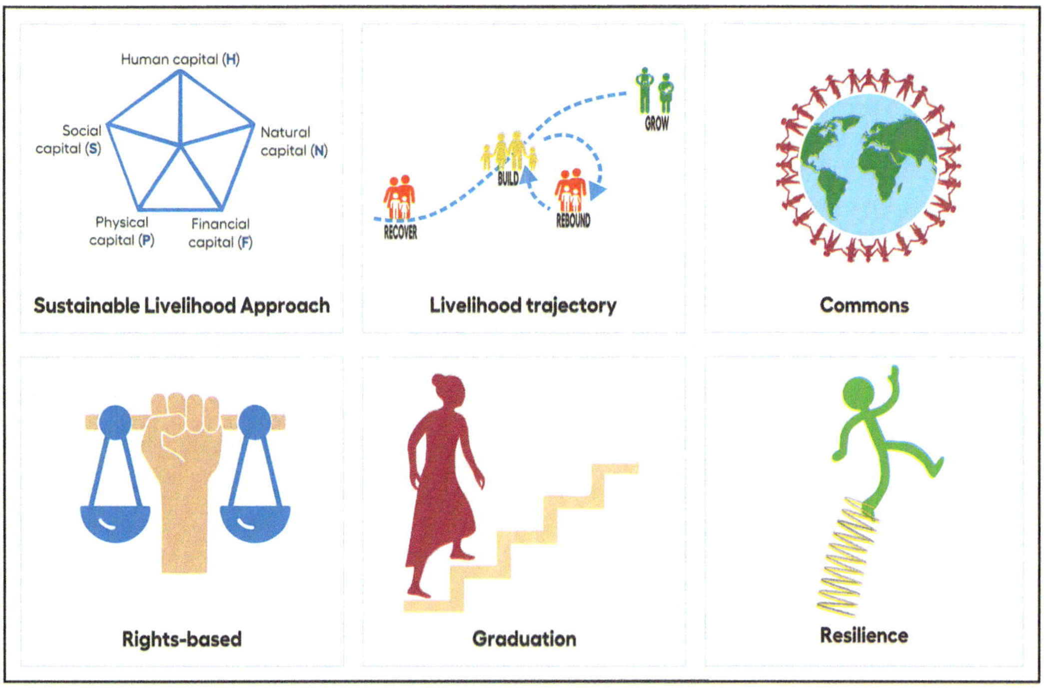

Fig. 1.2 The six prominent rural livelihood approaches

1.3.1 Sustainable Livelihoods Approach

In 1998, IDS published the working paper on sustainable rural livelihoods by linking the five elements of vulnerability context, household resources, policies and institutions, livelihood strategies, and livelihood outcomes (Scoones, 1998). This framework hinges on the five core principles of being people-centred, holistic, dynamic, building on strengths, macro-micro links, and sustainability (DFID, 1999). It has four main components. Households based on their vulnerability context, asset mix (human, social, physical, natural, and financial), and the prevailing policies, institutions, and processes, formulate their livelihood strategy to achieve corresponding livelihood outcomes which in turn again impact their asset pentagon and vulnerability context. This framework helped unpack the complexities of human livelihoods by integrating micro and macro elements such as the household asset pentagon, vulnerability context, and the policy and institutional setting (Solesbury, 2003). It was adopted by various global funding and research institutions and the early experiences and adaptations were compiled and integrated to develop the DFID sustainable livelihoods approach (Ashley & Carney, 1999; DFID, 1999). It received enthusiastic political support and instilled new energy and enthusiasm into international development. It was used extensively by bilateral and multilateral funding agencies and large NGOs, in their developmental approaches and aid for the global south. Various organizations such as UNDP, CARE, FAO, and others adopted this approach enthusiastically and created their own variants of livelihood approaches (Carney, 2002; Hussein, 2002; Scoones, 2009). The DFID SRL approach is credited with shifting the developmental thinking from funding sectoral interventions to household poverty alleviation (Mensah, 2012).

1.3.2 Commons and Livelihoods

Traditional rural livelihoods are intimately linked to natural resources such as grasslands, wetlands, forests, rivers, seas, etc., collectively known as common pool resources (CPRs). These resources share that the property of both public and private goods as their access is non-excludable while their consumption is rivalrous. Hence, as large number of people have access to them and they have a finite supply, they can get over-extracted and degraded. Hardin (1968) in his influential article titled, 'tragedy of the commons' argued that users would tend to maximize their individual use of this resource, resulting in its overuse and eventually impacting its sustainability. CPRs are susceptible to overuse as they create incentives for users to ignore the negative externality of their actions which are borne by the group which has to protect, conserve, and manage the resource. Ostrom (1990) in her seminal work on community-based governance of environmental commons challenged this theory that CPRs when managed by communities, would invariably result in their depletion and hence the only viable option was either state control or privatization. She demonstrated through rich empirical evidence from around the world that self-organized communities can effectively manage the commons with collective action (Frischmann, 2013). This framework postulates eight design principles associated with successful governance of the commons by local collective action (Ostrom, 2000). She also developed a general framework for analysing the sustainability of the socio-ecological systems (SES) (Ostrom, 2009). Through her work, Ostrom demonstrated that local communities could develop rules and enforcement mechanisms to effectively conserve and use their natural resources thereby supporting their livelihood and food security. Her work inspired a whole generation of researchers from diverse disciplines and helped create a body of knowledge on the commons.

1.3.3 Livelihood Trajectory

Livelihood trajectory refers to the changing ways in which households construct livelihood over time while responding to shocks, stresses, and opportunities (Bagchi et al., 1998). Insights into these trajectories are gained through temporal snapshots-based assessments and life histories that trace these household pathways (Ellis, 2000). They illuminate not only 'what' has changed but also explain 'why' and provide deep insights into the lived realities of the poor (Kothari & Hulme, 2004). Based on learnings from multi-country poverty dynamics trials, Krishna (2006, 2007) found that while the pathways out of poverty were diverse, death and illness of the household head contributed the most to the descent into poverty. This was followed by several studies on livelihood trajectories notably Dorward et al. (2009) and Mushongah (2009), who suggest a classification of the livelihood strategies pursued by the poor into four broad types. The first is 'hanging in' where households struggle to maintain their livelihood levels in the face of adverse conditions, the second is 'stepping up' or scaling up their existing activities, third is 'stepping out' where households use the accumulated assets to diversify into new activities, and the fourth is 'dropping out' due to chronic poverty or ill health. These studies bring into focus the concepts of chronic and transient poverty and also on the pathways into and out of poverty. Davis and Baulch (2011) integrated life-history interviews within a much larger quantitative panel survey to assess socio-economic mobility and life trajectories. Their analysis provided greater clarity on concepts such as transition matrix, life trajectory patterns, and drivers of poverty dynamics while elaborating on the challenges faced in integrating quantitative and qualitative data and how to address them. The livelihood trajectory approach helps to identify the shocks and stresses that impact livelihoods and the factors that enhance or impede the vulnerability and resilience of the households (Sallu et al., 2010). These studies play a key role in informing future poverty policies—one to block descent and the other to accelerate ascent (Krishna, 2007).

1.3.4 Rights-Based Approach

A human rights-based approach is defined as *a conceptual framework for the process of human development that is normatively based on international human rights standards and operationally directed to promoting and protecting human rights* (Broberg & Sano, 2018). This approach from the initial focus on civil and political rights, embraced economic, social, and cultural rights starting from the World Conference on Human Rights in Vienna in 1993, and then at the 1995 World Social Development Summit at Copenhagen (de Silva, 2013). During the latter half of the twentieth century, the development paradigm was that governments, and then the market would provide for the basic needs of the poor. But over the decades the deeper, structural problems of social justice, inequality, and discrimination persisted (Offenheiser & Holcombe, 2003). The focus of the rights-based approach was to identify and dismantle the key systemic barriers that prevented people from exercising their rights, building capabilities, and accessing opportunities (Offenheiser & Holcombe, 2003). In applying this approach in development, the focus is less on service delivery and general capacity building, and more on enabling the duty bearer that is the state to respond to the claims from the recipients of development assistance who are the right holders (Broberg & Sano, 2018).

1.3.5 Graduation Approach

The run-up to the millennium development goals saw impressive gains in reducing poverty. However uneven progress meant that the remaining global poverty was geographically concentrated with the added challenge of poverty depth. The Bangladesh Rural Advancement Committee (BRAC), a civil society organization in Bangladesh founded by Fazle Hasan Abed observed that the conventional anti-poverty initiatives while benefiting the poor were unable to sustainably impact the ultra-poor who continued to stagnate. The ultra-poor have little or no asset base, are highly vulnerable to shocks and chronically food insecure, with labour work being their only source of income (Halder & Mosley, 2004). They realized the need to recognize the ultra-poor as a separate category different from the poor, and to specially tailor-make a livelihood approach for them (Hashemi & De Montesquiou, 2011; De Montesquiou et al., 2014). Building on experience gained from practice, the graduation approach was designed comprising a combination of social security, livelihood development, and microfinance interventions coupled with market analysis and regular coaching. The term 'graduation' refers to ultra-poor households moving out of safety net programmes and 'graduating' into income-earning activities and sustainable livelihoods without external subsidies (El-Zoghbi et al., 2009). This approach involves a comprehensive, time-bound, and sequenced set of interventions that aim to graduate people from ultra-poverty to sustainable livelihoods. This approach was piloted in eight countries to test its efficacy and proof of concept. Following successful outcomes in this multi-country trial, the BRAC graduation approach has started gaining prominence in the developmental agenda (De Montesquiou et al., 2014). This approach directly feeds into the SDG goal of ending poverty in all forms everywhere as it specifically focussed on the vulnerable and those that are left behind. Many countries have now initiated pilots using the BRAC approach or are scaling up the pilots into full-fledged programmes.

1.3.6 Resilience Framework

Resilience is described as '*the capacity of people or systems to cope with stresses and shocks by anticipating them, preparing for them, responding to them, and recovering from them*' (Pain & Levine, 2012). In that sense, it is similar to an extension or rebranding of the sustainable rural livelihood approach (Carr, 2020). The relevance of resilience has grown across disciplines with the breadth of its application covering climate change, economic volatility, political instability, water security, safety nets, sustainable development, humanitarian aid, and others (Pain & Levine, 2012; Barrett & Constas, 2014). In the context of rural livelihoods, after graduating to a non-poor status, how do we safeguard the livelihoods from the risks posed by man-made and environmental shocks? Resilience is about enhancing the capacity to proactively and positively manage this change in ways that contribute to a world without poverty. How do households respond to changes in the external environment? The Oxfam framework for resilient development describes three types of resilience capacities: absorptive, adaptive, and transformative (Jeans et al., 2016). These three capacities need to be enhanced to achieve resilient development outcomes. Without resilience we run the danger of reversal of the developmental gains and only with resilience can we create more lasting solutions.

1.4 Patterns in Development Studies

Development thinking and livelihood approaches feed into the inter-disciplinary field of development studies. This field seeks to understand the various aspects of societal change, particularly in the context of developing countries, and aims at providing solutions to societal problems (Opschoor et al., 2005). While not being a discipline in itself, development studies synthesize the approaches and insights from several other disciplines. It is context specific and aims to take into account the specificities of different societies covering a wide range of topics such as poverty, environment, gender equity, human development, sustainable development, etc. which are constantly evolving (Opschoor et al., 2005). Traditionally, development studies were an asymmetric business, with largely the northern social scientists studying the south (Opschoor et al., 2005). This trend is gradually changing with the growth of home-grown institutes in the global south and local researchers contributing to the development debate. Mönks et al. (2017), studied the relevance of development studies, and found that it is thriving and expanding with growing demand from developing countries. They also report a proliferation of new themes such as health, water, sanitation, impact evaluation, resilience, tourism, cities, etc., with some themes gaining relevance such as climate change, sustainable development, globalization, governance, urban space, etc. Development studies as a research field are also thriving with both the number of journal titles and their impact factor growing (Mönks et al., 2017). As development studies expand to new themes there is a growing emphasis on research that informs policy and is practical at the same time. Hence, while livelihood approaches continue to inform mainstream development studies on poverty and livelihoods, there is an opportunity as well to impact the new domains of access, equity, disasters, urban spaces, resilience, climate change, governance, sustainable development, etc.

1.5 Linking Livelihoods with Science, Policy, and Practice

Despite such diverse academic insights available in the inter-disciplinary field of rural livelihoods, we submit that this knowledge has not been able to adequately inform the knowledge action pathway. There is clearly a need for a stronger connect between the rural livelihood approaches and their uptake in science, policy, and practice. Ironically one gap area that remains is that even poverty studies and livelihood research are not adequately informed by these livelihood approaches. Researchers rely on household schedules and participatory approaches to answer their research questions instead of feeding this data into the rural livelihood frameworks to gain new insights. We address this in Part 3 of this book, where we explore avenues to integrate these livelihood approaches in scientific studies and in policy and development. Livelihood development projects often do not build on these livelihood approaches, but instead rely on generalistic project management frameworks such as the logical framework approach, results framework, theory of change, etc. We opine that poverty and rural livelihood projects can instead build on these livelihood approaches to better inform project management frameworks so that they are based on a solid knowledge foundation.

1.6 Livelihoods, Development, and Resilience

Livelihood development, ending poverty, and resilience to shocks and stresses are influenced by both macro and micro factors. At the macro level, state led-development and market-led structural transformation impact the policies, programmes, and opportunities available at the local level. While the asset base, capabilities, and freedoms at the micro-level determine the extent to which the households are able to access and take advantage of these opportunities, understanding this macro-micro linkage between development, livelihoods, and resilience is the challenging part of sustainable development. For a holistic understanding of SDG 1 of ending poverty in all its forms everywhere, it is vital to combine the macro, comprising of transforming structures, processes, and institutions, with the micro, that embraces the vulnerability context, asset pentagon, livelihood strategy, and resilience of the livelihood outcomes (Bohle, 2009). We discuss the macro aspects in Part I of the book and the micro aspects in Part II, whereas Part III of this book focusses on how the livelihood approaches can be integrated, how they can better inform the macro perspectives, and how to strengthen their uptake in science, policy, and practice. While teaching and learning rural livelihoods, the challenge lies in operating at the intersection of academic rigour and applicability in practice. We provide insights on how to enhance the applicability of the rural livelihood approaches in various real-life scenarios. We encourage policymakers and practitioners to explore ways to integrate these livelihood approaches in their conceptual 'theory of change'. It is hoped that this book will bring about a clearer understanding of the various rural livelihood approaches and their application in practice to broaden their uptake in livelihood research, policy, and development.

References

Alkire, S., & Santos, M. E. (2014). Measuring acute poverty in the developing world: Robustness and scope of the multidimensional poverty index. *World Development, 59*, 251–274.

Ashley, C., & Carney, D. (1999). Sustainable livelihoods: Lessons from early experience (7). 1 : Department for International Development.

Bagchi, D. K., Blaikie, P., Cameron, J., Chattopadhyay, M., Gyawali, N., & Seddon, D. (1998). Conceptual and methodological challenges in the study of livelihood trajectories: Case-studies in eastern India and Western Nepal. *Journal of International Development, 10*(4), 453–468.

Barrett, C. B., & Constas, M. A. (2014). Toward a theory of resilience for international development applications. *Proceedings of the National Academy of Sciences, 111*(40), 14625–14630.

Bohle, H. G. (2009). Sustainable livelihood security. Evolution and application. In H. G. Brauch et al. (Eds.), *Facing global environmental change. Hexagon series on human and environmental security and peace* (Vol. 4). Springer. https://doi.org/10.1007/978-3-540-68488-6_36

Broberg, M., & Sano, H. O. (2018). Strengths and weaknesses in a human rights-based approach to international development–an analysis of a rights-based approach to development assistance based on practical experiences. *The International Journal of Human Rights, 22*(5), 664–680.

Carney, D. (2002). *Sustainable livelihoods approaches: Progress and possibilities for change*. Department for International Development.

Carr, E. R. (2020). Resilient livelihoods in an era of global transformation. *Global Environmental Change, 64*, 102155.

Chambers, R. (1997). *Whose reality counts*. Intermediate technology publications.

Chambers, R. (2008). PRA, PLA and pluralism: Practice and theory. *The Sage handbook of action research. Participative inquiry and practice*, 2, 297–318.

Chambers, R. and Conway, G. (1992) Sustainable rural livelihoods: Practical concepts for the 21st century. IDS Discussion Paper 296, Institute of Development Studies, Brighton.

Davis, P., & Baulch, B. (2011). Parallel realities: Exploring poverty dynamics using mixed methods in rural Bangladesh. *The Journal of Development Studies, 47*(1), 118–142.

de Haan, L., & Zoomers, A. (2005). Exploring the frontier of livelihoods research. *Development and Change, 36*(1), 27–47.

De Montesquiou, A., Sheldon, T., Degiovanni, F., & Hashemi, S. (2014). *From extreme poverty to sustainable livelihoods: A technical guide to the graduation approach*. CGAP and Ford Foundation.

de Silva, T. E. (2013). Principles in international development: Sustainable livelihoods and human rights-based approaches. Division of Economic and Labor Research Office of Trade and Labor Affairs U.S. Department of Labor, USA.

Deka, A. (2012). Local and global poverty: Insights using a rights-based approach. *Advances in Social Work, 13*(2), 471–483.

DFID. (1999). *Sustainable livelihoods guidance sheets, numbers 1-10*. Department for International Development.

Dorward, A., Anderson, S., Bernal, Y. N., Vera, E. S., Rushton, J., Pattison, J., & Paz, R. (2009). Hanging in, stepping up and stepping out: Livelihood aspirations and strategies of the poor. *Development in Practice, 19*(2), 240–247.

Elder, M., & Olsen, S. H. (2019). The Design of Environmental Priorities in the SDGs. *Global Policy, 10*, 70–82.

Ellis, F. (2000). *Rural livelihoods and diversity in developing countries*. Oxford University Press.

El-Zoghbi, M., Montesquiou, A. D., & Hashemi, S. (2009). Creating pathways for the poorest: Early lessons on implementing the graduation model. CGAP brief. Retrieved on 29th January, 2021 at 1:00 pm from https://www.cgap.org/sites/default/files/CGAP-Brief-Creating-Pathways-for-the-Poorest-Early-Lessons-on-Implementing-the-Graduation-Model-Dec-2009.pdf

Frischmann, B. M. (2013). Two enduring lessons from Elinor Ostrom. *Journal of Institutional Economics, 9*, 387–406.

Halder, S. R., & Mosley, P. (2004). Working with the ultra-poor: Learning from BRAC experiences. *Journal of International Development, 16*(3), 387–406.

Hardin, G. (1968). Tragedy of the commons. *Science, 162*, 1243–1248.

Hashemi, S. M., & De Montesquiou, A. (2011). Reaching the poorest: Lessons from the graduation model. *Focus Note, 69*(1), 1–15. https://www.tandfonline.com. https://doi.org/10.1080/02681102.2016.1233856

Hussein, K. (2002). *Livelihoods approaches compared: A multi-agency review of current practice*. DFID and ODI.

Jeans, H., Thomas, S., & Castillo, G. (2016). *The future is a choice: The Oxfam framework and guidance for resilient development*. Oxfam International.

Kothari, U., & Hulme, D. (2004). *Narratives, stories and tales: Understanding poverty dynamics through life histories. GPRG-WPS-011* (p. 41). ESRC Global Policy Research Group.

Krishna, A. (2006). Pathways out of and into poverty in 36 villages of Andhra Pradesh, India. *World Development, 34*(2), 271–288.

Krishna, A. (2007). For reducing poverty faster: Target reasons before people. *World Development, 35*(11), 1947–1960.

Mensah, E. (2012). The sustainable livelihood framework: A reconstruction. *The Development Revew Beyond Research, 1*(1), 7–25.

Mönks, J., Carbonnier, G., Mellet, A., & De Haan, L. (2017). Towards a renewed vision of development studies. *International Development Policy| Revue internationale de politique de développement*, 8(8.1).

Mushongah, J. (2009). *Rethinking vulnerability: Livelihood change in southern Zimbabwe, 1986–2006. Ph.D. dissertation*. University of Sussex.

Offenheiser, R. C., & Holcombe, S. H. (2003). Challenges and opportunities in implementing a rights-based approach to development: An Oxfam America perspective. *Nonprofit and Voluntary Sector Quarterly, 32*(2), 268–301.

Opschoor, H., Forster, J., Jolly, R., & Mönks, J. (2005). *Development studies, accreditation and EADI. A vision paper presented to the EADI Executive Committee*. European Association of Development Research and Training Institutes (EADI). https://www.eadi.org/typo3/fileadmin/Accreditation/Documents/EADI_Vision_paper_final.pdf Accessed 29 January 2021

Ostrom, E. (1990). *Governing the commons: The evolution of institutions for collective action*. Cambridge University Press.

Ostrom, E. (2000). Collective action and the evolution of social norms. *Journal of Economic Perspectives, 14*(3), 137–158.

Ostrom, E. (2009). A general framework for analyzing sustainability of social-ecological systems. *Science, 325*(5939), 419–422.

Pain, A., & Levine, S. (2012). A conceptual analysis of livelihoods and resilience: addressing the insecurity of agency. ODI.

Sallu, S. M., Twyman, C., & Stringer, L. C. (2010). Resilient or vulnerable livelihoods? Assessing livelihood dynamics and trajectories in rural Botswana. *Ecology and Society, 15*, 4.

Scoones, I. (1998) Sustainable rural livelihoods: A framework for analysis. Working Paper 72, Institute for Development Studies, Brighton, UK.

Scoones, I. (2009). Livelihoods perspectives and rural development. *The Journal of Peasant Studies, 36*(1), 171–196.

Sen, A. (1981). Ingredients of famine analysis: Availability and entitlements. *The Quarterly Journal of Economics, 96*(3), 433–464.

Sen, A. (1982). *Choice, welfare and measurement*. Blackwell. and Cambridge Mass: MIT Press.

Sen, A. (1985). Well-being, agency and freedom: The Dewey lectures 1984. *Journal of Philosophy, 82*, 4.

Sen, A. (1993). Capability and Well-being. In A. K. Sen & M. Nussbaum (Eds.), *The quality of life* (pp. 30–53). Oxford University Press.

Solesbury, W. (2003). Sustainable livelihoods: a case study of the evolution of DIFD policy. Overseas Development Institute, Working Paper 217, 1–36.

Chapter 2
Structural Transformation and Rural Livelihoods

Abstract This chapter paints the background or context of the book detailing the macro-level structural transformation underway and its impacts on rural livelihood diversification. It provides an overview of the restructuring of the economy and the different trajectories taken by the countries in the global south based on their political and socio-economic realities. During their economic take-off, developed countries were able to diversify away from agriculture into industry and services, resulting in rapid growth in overall productivity and incomes. While countries in South Asia and sub-Saharan Africa considered less advanced failed to achieve a similar transformation and have remained trapped in lower levels of incomes, with agriculture still contributing significantly to employment. In many developing countries, a rural transformation is underway—social equations are changing, connectivity and mobility have expanded, urban-rural boundaries are blurring, labour markets are integrating, and cash incomes are becoming important. Smallholder farmers are yet to fully benefit from livelihood diversification. Provisioning of full and productive employment and decent work for all is at the heart of rural transformation. We opine that a restructuring of the economy coupled with effective redistributive programmes can result in rapid poverty reduction and an inclusive transformation.

Keywords Economic growth · Employment · Structural change · Industrialization · Urbanization · Rural non-farm · Smallholder farmers · Rural development · Agriculture · Poverty reduction

2.1 Economic Growth and Employment

In this chapter, we enlarge the canvas and take a bird's eye view of rural livelihood patterns. While the micro-assessment of rural livelihoods at the household and community level covering farm and non-farm livelihoods is the focus of subsequent chapters, here we take a step back and look at the regional picture with a special focus on Asia and SSA. We use the lens of structural reforms to assess the macro picture, a process that has deeply impacted livelihoods and poverty reduction worldwide (IFAD, 2016). The idea behind Goal 8 of the SDGs is 'economic growth that is sustained, inclusive, sustainable, and results in full and productive employment and decent work for all'. Job creation has now become the topmost priority of governments worldwide and figures prominently on the political agenda and in national policymaking. All countries aspire for economic growth that generates decent and well-paid jobs that will lead to poverty reduction (Majid, 2019).

S. Tambe, *Teaching and Learning Rural Livelihoods*, Sustainable Development Goals Series,
https://doi.org/10.1007/978-3-030-90491-3_2

2.2 Structural Change of the Economy

Structural change is the process of reallocation of labour across economic sectors with different levels of labour productivity and is both an outcome of and a contributor to the growth process in an economy (Lewis, 1954; Ghose, 2020). IFAD (2016) defines it as *the transition of an economy from low productivity and labour-intensive economic activities to higher productivity and skill-intensive activities*. As the workforce transitions from labour-intensive to skill-intensive activities, it contributes positively to growth (UN Habitat, 2016; Ghose, 2020). IFAD (2016) highlights that 'The driving force behind the structural transformation is the large differential in productivity in the modern sector, which is dominated by manufacturing and services. It involves a declining contribution of agriculture in GDP and employment, rural-urban migration, accelerating urbanization and a rise in the modern industrial and service economy'. Structural transformation is then a principal route for improvement in employment conditions in low-income countries, where the employment condition manifests itself in the form of a prevalence of low-income and low-productivity jobs (Ghose, 2020). Structural transformation is necessary not only for high productivity and incomes but also in creating an economy that is diversified and resilient to poverty and external shocks (UN Habitat, 2016). In this process of structural change, countries transition from low to high-income with the contributing factors being rapid urbanization, income growth, poverty reduction, and lower birth rates (Barrett et al., 2017).

In the global north, before the industrial revolution, agriculture was the mainstay of the economy and employment (Box 2.1). With industry becoming productive, the share of agriculture in GDP fell faster than its share in employment. Farm incomes and agricultural productivity started lagging behind the other sectors of the economy. As the demand for labour grew in the non-agriculture sector, labour was pulled out of agriculture (Binswanger-Mkhize, 2013). The rapidly growing manufacturing sector absorbed the surplus labour from rural areas. A special characteristic of manufacturing is that even low-skilled labour can graduate to productive, high paying jobs which further enhances growth and work conditions (Kaldor, 1967; Ghose, 2020). This pull factor exerted by higher labour productivity in the manufacturing sector resulted in a large-scale rural-urban migration leading to a rapid decline in the share of the labour force in agriculture (Fig. 2.1). As a consequence, there was a consolidation of rural land holdings, growth in agricultural productivity, and the difference in the labour productivity between the primary and secondary sectors narrowed. These structural changes resulted in a rapid rise in the agriculture value added per worker thereby contributing to inclusive growth (Fig. 2.2). Countries that could not attain this ideal structural transformation, as non-farming jobs could not be created at a fast face, have higher employment in the agriculture sector associated with higher multidimensional poverty (Fig. 2.3).

Box 2.1 The Global North-South Divide

The global north-south divide is a socio-economic division of the world that gained popularity in the late twentieth century. Global north or south does not imply a strict geographical classification, as there are exceptions to this. For example, Australia is in the global south geographically but it is considered as a part of the global north in terms of its socio-economic status. The global north-south divide is the gap between the countries in terms of development and wealth. Being categorized as 'north' implies development as opposed to 'south', which implies a lack thereof. In the 1980s, the Brandt line was developed as a visual model for the north-south divide. It was proposed by the former chancellor of West Germany, Willy Brandt and hence the name. As per this model, the richer countries are almost all located in the northern hemisphere, whereas the poor countries, in general, are located in the tropical region and the southern hemisphere. With time, the notion of two distinct economic spheres has been challenged.

(continued)

Box 2.1 (continued)

Globalization, structural reforms, and regional cooperation in Asia have led to a more equitable distribution of global economic power. Relatively poor countries in the global south have experienced significant economic growth and social development in the recent past, making the world much more complex than what the Brandt line depicts. Hence, the use of 'global north' and 'global south' in the present discourse is generally used to refer to the richer or poorer communities both within and between countries in both the hemispheres.

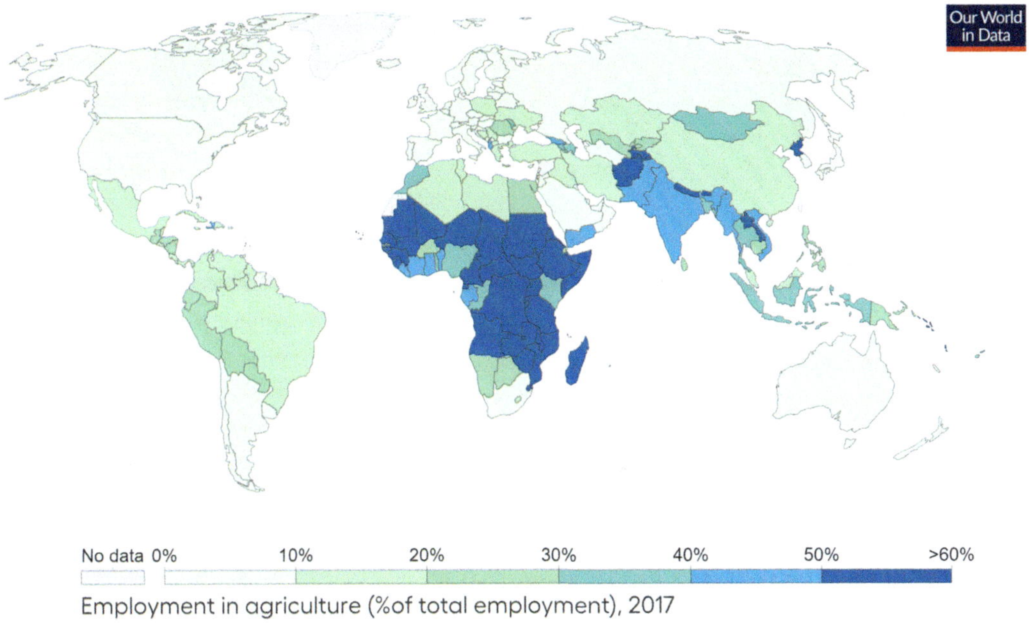

Fig. 2.1 Share of labour force employed in agriculture

Historically industrialization and rising agricultural productivity have gone hand in hand. Under structural transformation, the contribution of agriculture to the overall economy declines, but this should not be construed as a lack of importance of the agricultural sector in economic development as it plays an important role through agribusiness and agro-industry (IFAD, 2016). This process of transformation of the economy thus profoundly impacts the rural landscape with its interrelationships between agriculture, rural non-farm economy, manufacturing, and services (IFAD, 2016).

2.2.1 Historical Pathway of Structural Change

In developing countries with low per capita incomes, before structural transformation, the proportion of workers engaged in the agriculture sector is typically 75%, while for developed countries with higher per capita income, it is less than 10% (IFAD, 2016). Strong rural-urban linkages facilitate this type of transformation and result in rapid poverty alleviation and inclusive rural development. However, when the modern economy and the non-farm sector are not growing fast enough, it will be unable to absorb the surplus rural labour force that will remain underemployed (Timmer, 2014).

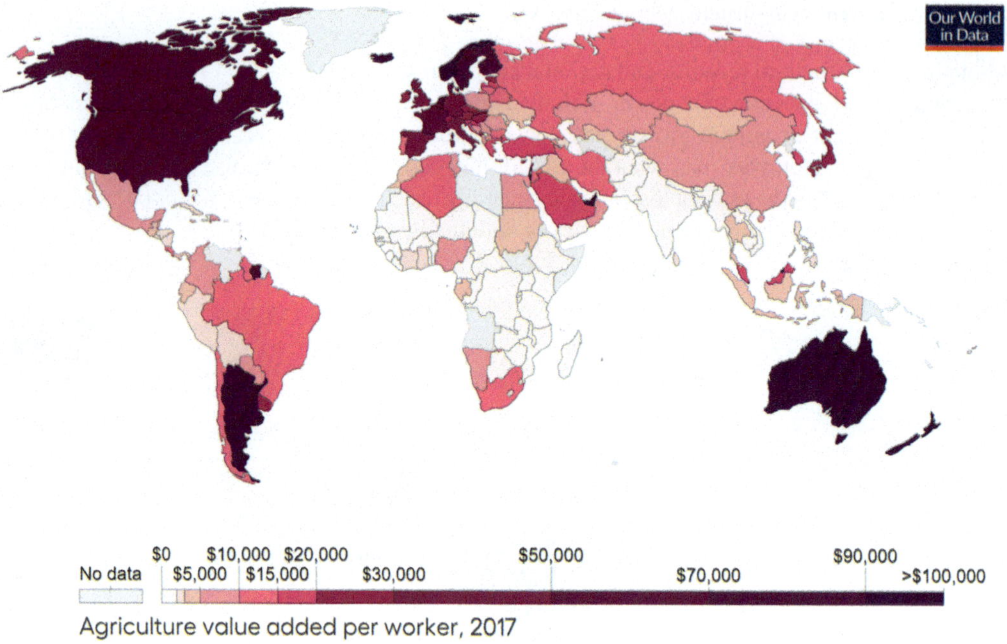

Fig. 2.2 Agriculture value added per worker

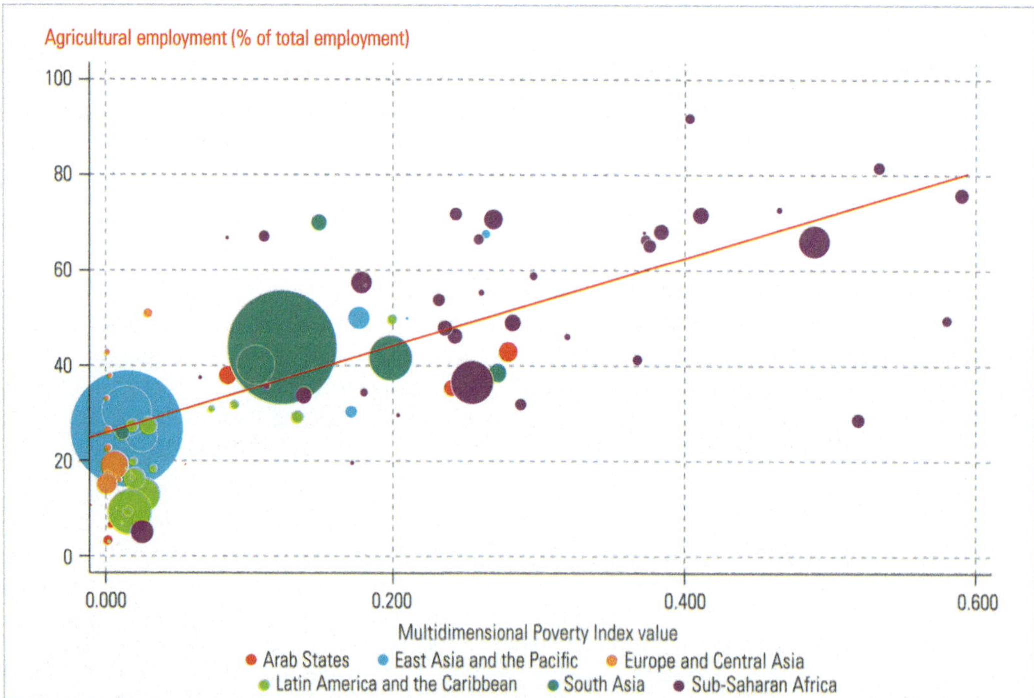

Fig. 2.3 Higher employment in the agricultural sector is associated with higher multidimensional poverty in sub-Saharan Africa (Source OPHI 2020)

Hence, to attain an ideal structural transformation, non-farming jobs need to be created at a faster pace. While in the global north, urbanization has attained an equilibrium, in the global south, rural is the new urban. By 2050, it is projected more than 60% of the world's population will reside in urban areas (UN Habitat, 2016).

The structural transitions in Great Britain followed by Western Europe and then by North America started in the nineteenth century. In Latin America, it started in the early twentieth century while in East Asia it began in the mid-twentieth century. Late developing countries in parts of South Asia, Southeast Asia, and sub-Saharan Africa (SSA) embarked on this process in the last decade of the twentieth century. This structural shift of the economy in many countries of East and Southeast Asia was transformative, driven by rapid growth in industry and modern services as demonstrated by China. The effect of structural transformation experienced by them was the predominant movement of labour to industry and modern services. By 2020, the contribution of agriculture to GDP in China had fallen to 7% and it employed only 25% of the workforce. The combination of a 'pull' factor from external economic development and 'push' forces from growing aspirations and local conditions resulted in diversification of rural livelihoods involving ownership of diverse livestock, non-farm employment, and outmigration. It was enabled by strong growth in manufacturing with the contribution of industry to the economy hovering at 46% since the early 1990s. Wang et al. (2011) report substantial changes in rural livelihoods in China driven by a combination of "pull" forces from external economic development, and 'push' forces from local areas, leading to a shift in rural household economic activities: household outmigration and depopulation of the countryside, changes in energy consumption, changes in land use, and ecological restoration. They report that in the 1970s the rural workforce was fully engaged in agriculture, but now more than 70% has transitioned to non-farm employment. The population growth has declined to nearly zero and the rural-urban migration has involved 220 million workers in the past two decades.

2.2.2 *Structural Transformation in Latin American Countries*

The Latin American countries underwent a rapid structural transformation in the second half of the twentieth century with agriculture losing its prominence with rapid industrialization, livelihood diversification, and growth in rural non-farm incomes (Box 2.2) (IFAD, 2016). Today Latin America is the world's most urbanized region with 80% of the population living in cities (UN Habitat, 2016). This rapid rural transformation resulted in different degrees of social inclusion across the region. Inequality, violence, and poor living conditions in the cities are cited as the main negative impacts of this rapid urbanization (UN Habitat, 2016). Globalization resulted in premature deindustrialization or phasing out of industries that could not compete with cheap imports, resulting in unemployment with labour having to search for other less productive avenues and needing social support programmes (IFAD, 2016).

Box 2.2 The Rural Non-Farm Sector

The rural non-farm (RNF) sector includes all economic activities like handicrafts, repair services, construction activities, transportation, trade, household, and non-household manufacturing, other than agriculture or farm-based economic activities in rural areas. RNF activities help in the diversification of household incomes and reduce the dependence of households on agriculture alone. It also helps in facilitating employment growth in rural areas, prevents rural to urban migration into the informal sector, and also acts as a coping mechanism for the rural

(continued)

Box 2.2 (continued)

masses when farming fails. Although RNF is an important tool in helping diversify the household incomes in rural areas, there are many inherent challenges to accessing the sector like education, gender, and issues of mobility. Apart from that, the quality, quantity, and reliability of infrastructure, regulatory restrictions, and lack of awareness also add to the challenges. Having said that, agriculture to the rural areas cannot be substituted entirely by RNF activities. It can at best act as an important supplementary measure to support the economy in the rural areas, especially in regions where the structural transformation is stunted and incomplete. In the context of rural areas, agriculture will continue to be the most important economic activity. The RNF sector has the potential to absorb the surplus labour not employed by the urban economies. This can happen only when efforts are made at the macro level to identify appropriate and effective institutional vehicles for the development of RNF policy and translating it into practice for creating employment opportunities.

2.2.3 Structural Transformation in Other Countries of Asia and SSA

In the other late developing countries of Asia and SSA, the structural transformation is rather incomplete due to the sluggish industrial growth. In India, the contribution of agriculture to GDP fell to 16% in 2020, but its share in employment is still sizeable at 41%. In SSA, like other late developing countries of Asia, the contribution of agriculture has declined to 15% of the total GDP but it still employs more than 50% of the total labour force (OECD-FAO, 2016). This gap between the share of agriculture to the economy and its contribution to employment is widening (Binswanger-Mkhize, 2013). Consequently, while there is a push from rural areas driven by fragmented land holdings, declining agricultural productivity, weak infrastructure, and unemployment, but there is a weak pull from urban areas to absorb this labour. The limited demand for jobs is in the informal sector and is insecure as it does not provision either social security, quality work conditions, or decent pay. As a result, while the contribution of agriculture to the economy is rapidly declining, but the share of the labour force in this low productive sector continues to be high resulting in unemployment and persistent poverty. The slow decline in the agricultural labour force is due to population growth and the relatively slow rate of urban-rural migration (Binswanger-Mkhize, 2013). In countries such as India with a weak manufacturing growth, commercialization in agriculture has benefitted many farmers, but it has also left out many small- holder or dry-land farmers. Unlike the successful Asian economies, the green revolution was not followed by an industrial revolution and rapid urbanization with economic activities shifting away from agriculture (UN Habitat, 2016). These transformations are characterized by a distinctly smaller role played by manufacturing, with the services and the rural non-farm sector contributing to employment. In SSA, there is little evidence of the green revolution as agricultural productivity has remained low and there was no industrial revolution either (UN Habitat, 2016). Hence, in these late developing countries, workers move primarily from the agriculture sector to the rural non-farm sector, rather than to the urban economy (Binswanger-Mkhize, 2013). As a result, rural areas in many of these countries in the global south are characterized by a persistence of small holder farming due to incomplete structural transformation resulting in jobless growth.

This trajectory of structural reforms demonstrated by these countries in Asia and SSA differed from the pattern of past transitions, wherein the workforce progressively shifted from agriculture to industry, and then to services. This historical transformation was driven by a green revolution and mechanization of farming which allowed the surplus labour and capital to flow to the modern sectors (Losch et al., 2012). This in turn triggered the process of urbanization where scattered agricultural activities transform to concentrated industrial ones resulting in higher incomes, better living conditions followed by stabilization in population. The manufacturing-led growth model resulted in rapid transfer of low-skilled labour from agriculture to manufacturing and services with rapid improvement in employment conditions (Ghose, 2020). In contrast, the services-led growth model is associated with the slow growth of jobs in manufacturing and services resulting in a slow improvement in employment conditions (Ghose, 2020). Thus these late developing countries of Asia and SSA could not create adequate productive jobs in the labour intensive manufacturing sector for its low-skilled population. The structural transition in these countries termed as premature services-led growth differs from the historical manufacturing-led transformation pathway as they are simultaneously confronted with incipient economic transition, unachieved demographic transition, and growing urbanization without industrialization. These countries faced challenges posed by a global open economy, high international competition, environmental degradation, and fast-growing populations. This experience from late developing countries underscores the point that 'moment in time' matters as opportunities, constraints, and power equations evolve (Losch et al., 2012).

2.3 Growth in the Rural Non-Farm Sector

For countries that missed the early mover advantage on using structural transformation as a means of absorbing the surplus labour from the rural areas in industry, what is the way forward? With urban areas not providing gainful employment opportunities, and with agriculture stagnating, rural poverty should have spiralled in these countries along with the rural-urban divide. Where has the surplus rural labour found employment and opportunities for increasing their income (Binswanger-Mkhize, 2013)? In these countries, the rural non-farm (RNF) sector is bridging this gap between subsistence farming and modern industrial and service sector jobs in towns and cities. The surplus labour is transitioning from agriculture into the RNF sector (Haggblade et al. 2007). Better access to education, health, basic amenities, roads, and mobile connectivity has also resulted in a blurring in the urban-rural boundaries. With growing aspirations and the need for cash incomes, rural areas are witnessing a diversification of livelihoods, with RNF incomes contributing sizeably to total household income. The RNF employment now constitutes a sizeable part of household income having grown to 37 percent in Africa, 47% in Latin America, and 51% in Asia (Haggblade et al. 2007). In the global south, structural transformation is underway resulting in rising growth in RNF incomes by diversification, while agricultural incomes continue to remain important (Davis et al., 2010). Income from farms and forests still comprises a sizeable portion of the total household income with environmental income share being higher for low-income households (Angelsen et al., 2014). However, the RNF jobs demand certain skills, mobility, and assets which the poor and other vulnerable groups in rural areas often lack (Haggblade et al. 2010; Nagler and Naudé 2014).

2.3.1 Growth in the Rural Non-Farm Sector in India

The rise in RNF sector needs to be looked upon as a diversification strategy and pluriactivity as it is in addition to subsistence farming and may also be agriculture distress driven (Guin, 2018). Most of the non-agricultural works are either casual labourers in the construction sector or self-employed as the availability of decent non-farm jobs is limited (Binswanger-Mkhize, 2013; Guin, 2018). The RNF employment increased more in regions with rapid urban growth as well as continued agricultural growth (Binswanger-Mkhize, 2013). Remote locations with poor agricultural endowments and forest fringe habitations are likely to be left out from these structural reforms as they are located outside the intersection of spillovers from the urban economy and agricultural growth. In India, those with higher

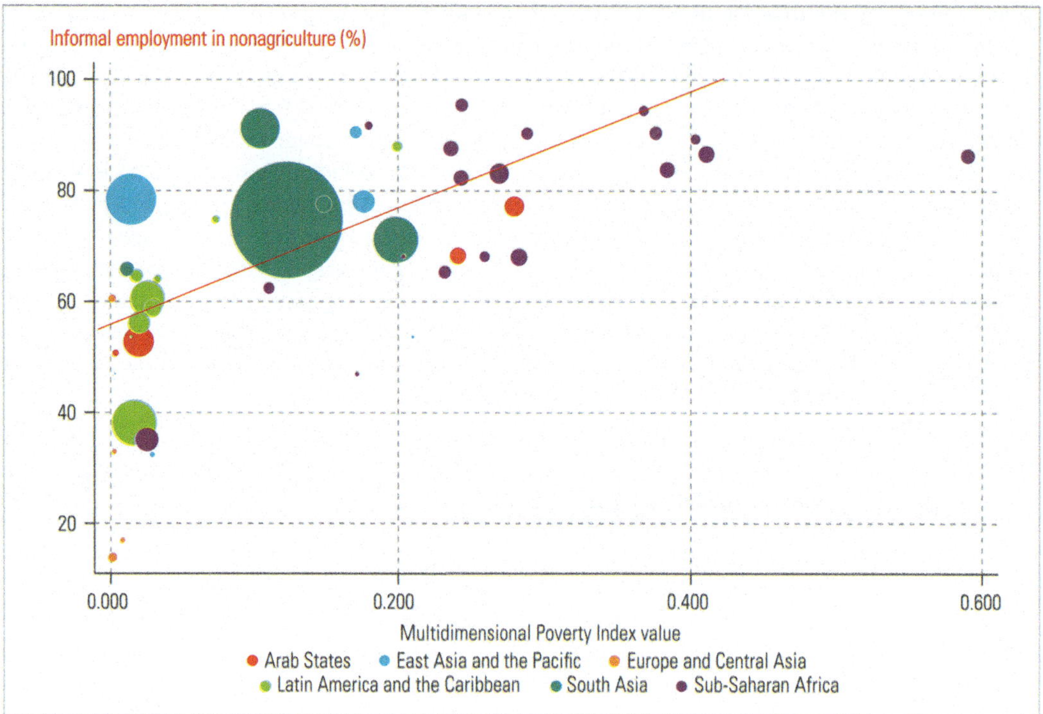

Fig. 2.4 A higher share of informal employment in non-agricultural employment is associated with higher multidimensional poverty (Source OPHI 2020)

levels of education and land ownership are able to access better paying RNF employment (Lanjouw and Shariff, 2004). Also, these jobs demand an inherent level of mobility and are more accessible to young men with some education than women. Hence, though the RNF employment provides an outlet for livelihood diversification it has both gender and educational attainment barriers. As a result, the poorest, landless, and women are excluded from RNF employment and are constrained to continue with farm labour work or casual labour work, which is unskilled, underpaid and has no avenues for growth. Hence, the evidence of the pro-poor impact of the RNF sector is mixed, with men having better opportunities than women. The social hierarchies, differential incomes, and discriminatory practises prevalent in the rural areas are also reflected in the RNF sector (Jodhka & Kumar, 2017). Consequently, agriculture will continue to feminize and will be dominated by small holder farmers (Binswanger-Mkhize, 2013). While RNF employment is desirable over being completely dependent on small holder farming, however, most of these enterprises are small and temporary and incapable of stimulating economic dynamism (Jodhka & Kumar, 2017). Hence, while the service sector has grown on the back of a skilled workforce, the challenge for India is to create well-paying and productive jobs on-scale in the non-farm sector, to attract the surplus, semi-skilled, rural workers. Globally a higher share of informal employment in the RNF sector is associated with higher multidimensional poverty (Fig. 2.4).

2.3.2 *Growth in the Rural Non-Farm Sector in SSA*

Before embarking on structural adjustment in the 1990s, SSA was known more for its civil wars, famines, and humanitarian crisis (Beegle et al., 2016). It was the only developing region where the MDG goal of halving extreme poverty by 2015 was not met. Over the last few decades, structural adjustment and market liberalization in sub-Saharan Africa triggered de-agrarianization and de-peasantization of the countryside (Bryceson, 2002). Ellis and Freeman (2004) studied east African countries and found that low household incomes in rural areas of all countries are associated with low land and livestock holdings, high reliance on food crop agriculture, and low monetization of the rural economy (Fig. 2.3). The challenge of persistent extreme poverty prevails due to weak structural transformation, as 82% of the poor are in rural areas, high population growth of 2.7% per year, and a high (69%) dependence on agriculture which is plagued by low productivity (Beegle et al., 2016; Barrett et al., 2017; Davis et al., 2010; Jayne et al., 2010). The annual cereal yields in SSA are still hovering at one metric ton per ha while those in the Asia Pacific are three times that. Better-off households are distinguished by virtuous spirals of accumulation typically involving diverse livestock ownership, engagement in non-farm self-employment, and diversity of on-farm and non-farm income sources. Loison (2015) reports that the poorest and vulnerable are unable to take advantage of non-farm opportunities, and it is the better-off smallholders having sufficient assets who succeed in livelihood diversification. Neves and Du Toit (2013) point out that this trajectory of livelihood transition of SSA of 'jobless de-agrarianization' is a result of lack of decent employment opportunities in the formal and informal sectors, and the problem of persistent rural poverty is less one of exclusion from the economic mainstream and more of an adverse engagement in the informal sector (Fig. 2.4).

2.4 Persistence of Smallholder Farmers

With economic progress, it was expected that like in the developed world the smallholders in the global south would also fade away (Rigg et al., 2016). However, in the late developing countries in Asia and Africa, the lack of opportunities in industry, growing rural population, and expansion in the RNF employment has resulted in a persistence of small holder farmers. A majority of the poor and hungry people in the world stay on small farms and eke out a livelihood from fragmented land holdings. As per an assessment by IFAD (2015), 'Small family farms dominate rural landscapes across the developing world and account for 80% of food produced in Asia and SSA while supporting livelihoods of up to 2.5 billion people'. These small holders can be categorized into two types, subsistence and business oriented (Hazell & Rahman, 2014). While the subsistence small holders have limited access to finance, use low inputs, and traditional methods in farming to eke out a livelihood, the business oriented have shifted to commercial farming and are succeeding in farming business often in combination with RNF employment (Hazell & Rahman, 2014). This intensification has also resulted in overuse of natural resources resulting in over grazing of pastures, depletion of groundwater, and forest degradation which has impacted the very sustainability of these livelihoods (IDFC Rural Development Network, 2013).

With the size of these farms being sub-optimal, how do the subsistence small holder farmers survive? Here we need to understand that the economics of farming and the economics of small holder are not the same (Rigg et al., 2016). Three main factors have aided in the survival of the small holders. Firstly, the political economy of the smallholders has ensured that there is a generous allocation of public funds in the form of public work programmes, food and cash transfers, subsidy on agricultural inputs, price support, and the like year after year. Governments in these countries have been demonstrating their commitment to the small holders by provisioning these safety nets and thereby subsidizing their livelihoods. Secondly, across parts of Asia and SSA which have not benefitted from fast-track industrialization, the small holder farmers have a higher reliance on the RNF sector. Livelihood diversification is pervasive and has ensured that small holders do not face an existential crisis as now farming is subsidiary to their living (Rigg et al., 2016). Thirdly, the nature of the RNF employment is on insecure terms with limited or negligible social benefits. Hence, it is a combination of insecure RNF employment, social safety nets, and subsistence farming that collectively provision a semblance of a secure livelihood for the smallholders. Therefore, though the farm size is sub-optimal and farming no longer occupies the central position it once did, the political economy of farming along with livelihood diversification, albeit on insecure terms ensures the persistence of smallholders farming in the global south (Rigg et al., 2016).

2.5 Structural Transformation, Rural Development, and Poverty Reduction

In the developing world, decent work deficits are the root cause behind persisting poverty (ILO, 2016). Here, the poor are caught in informal and insecure jobs with little or no social protection, usually in rural areas in the agriculture sector (Fig. 2.4). Tackling poverty sustainably will entail provisioning jobs that are secure, provide fair wages, decent working conditions with social security (Fig. 2.5). Efforts to fight poverty are destined to fail unless decent and better paid jobs are created for the poor and the unemployed. Rada and Von Arnim (2012) labelled the failure to create decent jobs in an expanding economy as *jobless growth* or *growth without development*. Hence to achieve inclusive

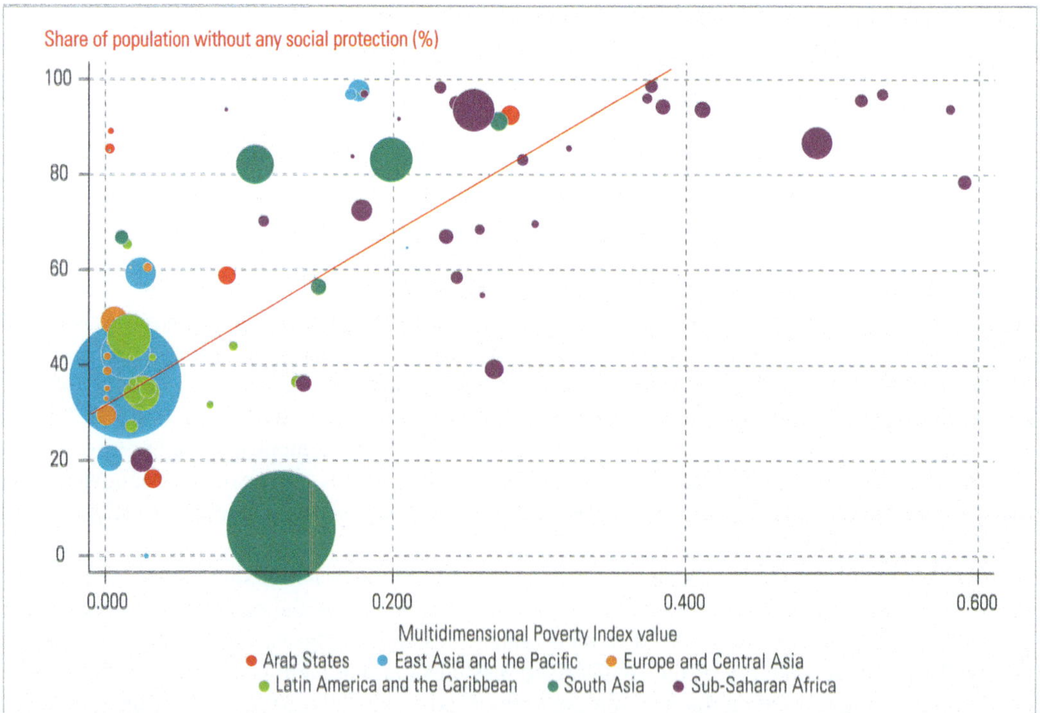

Fig. 2.5 In countries with high multidimensional poverty a large share of the population lacks any social protection (Source OPHI 2020)

growth, countries need to not only enable structural transformation but also take proactive steps towards rural transformation and poverty reduction (IFAD, 2016). Structural transformation as discussed earlier is the process of release of surplus labour and capital from agriculture to the more productive industrial and services sectors. Rural development is understood as the rise in agriculture productivity, expanded RNF employment opportunities, better access to infrastructure, basic amenities and services, and effective decentralization with a role in decision making (IFAD, 2016). It has the potential to enhance overall human wellbeing characterized by growth in life expectancy, improved education and health attainments, and access to basic amenities such as water, sanitation, housing, etc. Countries that have achieved a high level of structural and rural transformation generally show significant poverty reduction as well. On the other hand, some fast transforming countries have not been able to transform in an inclusive manner (IFAD, 2016). Hence, while structural transformation is necessary for inclusive rural transformation and sustainable development, it is not a sufficient condition (IFAD, 2016). Structural transformation by itself does not automatically result in inclusive growth and countries need to enact proactive policies and programmes to enable the rural people to benefit from this process and safeguard them from the challenges posed by it. Growth in agriculture reduces poverty faster than growth in other sectors and the advantage is largest for the poorest in society (Christiaensen & Martin, 2018). This finding especially underscores the role that agriculture development can play in alleviating poverty and fighting malnutrition especially in SSA and Asia where the world's poverty is increasingly concentrating (Christiaensen & Martin, 2018). Parikh et al. (2013) emphasize the importance of robust agriculture development for rapid and inclusive economic growth, which can be achieved by a growth in total factor productivity and expansion of the net irrigated area. As structural transformation unfolds governments need to promote agricultural growth, strengthen the viability of smallholder farming, and accelerate rural development. Inclusive rural transformation is at the heart of attaining sustainable development, reducing disparities and inequality, and fostering a peaceful and prosperous society.

Exercises

1. While structural transformation may be a central element and a necessary condition for poverty reduction but it is not sufficient (IFAD, 2016). What policy and programmatic actions would you propose to ensure inclusive rural transformation where marginalized groups, women and youth, small holder farmers, small and medium enterprises, and indigenous people are not left behind?
2. The late developing countries of Asia and SSA missed the manufacturing-led transformation pathway and relied more on the rural non-farm sector for growth and jobs. Discuss the national strategy they should adopt to achieve SDG Goal 1 of ending poverty in all its forms everywhere.
3. The contribution of agriculture to employment in developing countries is more than 50%, while in developed countries it is below 10%. Labour productivity in agriculture is typically lower than that in manufacturing and services. Hence, the contribution of agriculture to the economy is always lower than its labour share. In this context, what strategy should the developing countries adopt to create a pull factor enabling the movement of small holder farmers to the more productive industry and services sector?
4. Analyse why the trajectory of structural reforms demonstrated by the late developing countries of Asia and SSA differed from the historical manufacturing-led transformation pathway, wherein the workforce progressively shifted from agriculture to industry and then to services?
5. In light of the patterns of jobless growth, urbanization without industrialization, and persistence of small holder farming experienced by the late developing countries of Asia and SSA, discuss how their incipient structural transformation will pan out over the next few decades?
6. The dynamics of socio-economic mobility of two Indian villages was studied by Wilson (2004) using a comparative approach. Discuss the drivers of structural change and their impacts on the livelihoods of the caste groups along with their coping and accumulation strategies in these two villages.

https://www.odi.org/sites/odi.org.uk/files/odi-assets/publications-opinion-files/2474.pdf

7. The structural transformation of India has been atypical compared to China (Binswanger-Mkhize, 2013). Its transformation is characterized by slow absorption of labour in the urban economy, growth in the rural labour force, a decline in the size of landholdings, and persistence of small holder farming. What are the reasons that can be attributed to this stunted structural change in the Indian economy?
   https://www.jstor.org/stable/23527235
8. The rural livelihood diversification in Nepal and its role in household wellbeing were studied by Gautam and Andersen (2016). They found that diversification was not associated with wellbeing per se but with whether the household was involved in 'high return' sectors such as salaried jobs and trade or 'low return' sectors such as wage labour. Analyse how the livelihood capitals of the households determine its access to the type of rural non-farm employment and what steps are needed to counter this inequality of access to new economic opportunities.
   https://doi.org/10.1016/j.jrurstud.2016.02.001
9. A study on the 'Social Dynamics of Non-Farm Economy in the Madhubani District of Bihar, India' was conducted by Jodhka and Kumar (2017). The study showed that the role of agriculture has declined socially and is no longer viewed as an aspirational occupation. Non-farm employment and migration outside the state have grown due to a push factor contributed by natural calamities, lack of irrigation infrastructure, and lack of industrial growth, etc. Discuss the different types of non-farm occupations and their linkage with education, caste, and gender.
   https://www.researchgate.net/publication/318493135_Non-farm_economy_in_Madhubani_Bihar_Social_dynamics_and_exclusionary_rural_transformations
10. Rigg et al. (2016) argue that 'With economic progress, it was expected that the smallholders would fade into history like has been the experience in much of the global North. However, smallholders in the global South have persisted even in the face of rapid and profound social and economic transformation'. Study their journal paper and analyse the reasons behind the persistence of small holder farming in Asia against the economic odds.
    https://www.sciencedirect.com/science/article/pii/S0743016715300395
11. Is there a relationship between structural transformation, poverty, and HDI? Draw the graph between the share of agriculture in employment versus variables such as extreme poverty, MPI and HDI. What are the patterns that emerge? Use the global datasets available online such as https://ourworldindata.org/, http://hdr.undp.org/, http://hdr.undp.org/en/2020-MPI and others.

References

Angelsen, A., Jagger, P., Babigumira, R., Belcher, B., Hogarth, N. J., Bauch, S., … Wunder, S. (2014). Environmental income and rural livelihoods: A global-comparative analysis. *World Development, 64*, S12–S28.

Barrett, C. B., Christiaensen, L., Sheahan, M., & Shimeles, A. (2017). *On the structural transformation of rural Africa*. The World Bank.

Beegle, K., Christiaensen, L., Dabalen, A., & Gaddis, I. (2016). *Poverty in a rising Africa*. Africa Poverty Report. The World Bank.

Binswanger-Mkhize, H. P. (2013). The stunted structural transformation of the Indian economy: Agriculture, manufacturing and the rural non-farm sector. *Economic and Political Weekly, 48*, 5–13.

Bryceson, D. F. (2002). The scramble in Africa: Reorienting rural livelihoods. *World Development, 30*(5), 725–739.

Christiaensen, L., & Martin, W. (2018). Agriculture, structural transformation and poverty reduction: Eight new insights. *World Development, 109*, 413–416.

Davis, B., Winters, P., Carletto, G., Covarrubias, K., Quiñones, E. J., Zezza, A., … DiGiuseppe, S. (2010). A cross-country comparison of rural income generating activities. *World Development, 38*(1), 48–63.

Ellis, F., & Freeman, H. A. (2004). Rural livelihoods and poverty reduction strategies in four African countries. *Journal of Development Studies, 40*(4), 1–30.

Gautam, Y., & Andersen, P. (2016). Rural livelihood diversification and household well-being: Insights from Humla, Nepal. *Journal of Rural Studies, 44*, 239–249.

Ghose, A. K. (2020). *Structural transformation of India's economy, IHD Working Paper Series, WP01/2020*. Institute for Human Development.

Guin, D. (2018). From large villages to small towns: A study of rural transformation in new census towns, India. *International Journal of Rural Management, 14*(2), 87–109.

Haggblade, S., Hazell, P. B. R., & Reardon, T. (Eds.). (2007). *Transforming the Rural Nonfarm Economy: Opportunities and Threats in the Developing World,* Johns Hopkins Univ Press, Baltimore.

Haggblade, S., Hazell, P., & Reardon, T. (2010). The rural non-farm economy: Prospects for growth and poverty reduction. *World development, 38*(10), 1429–1441.

Hazell, P. B., & Rahman, A. (Eds.). (2014). *New directions for smallholder agriculture*. OUP.

IDFC Rural Development Network. (2013). *India rural development report 2012–13*. Orient BlackSwan.

IFAD. (International Fund for Agricultural Development) (2015). *Getting to work: Financing a new agenda for rural transformation*. IFAD.

IFAD. (International Fund for Agricultural Development) (2016). *Rural development report 2016. Fostering inclusive rural transformation*. IFAD.

ILO. (2016). *Transforming jobs to end poverty*. International Labour Organization, World Employment Social Outlook (WESO).

Jayne, T. S., Mather, D., & Mghenyi, E. (2010). Principal challenges confronting smallholder agriculture in sub-Saharan Africa. *World Development, 38*(10), 1384–1398.

Jodhka, S., & Kumar, A. (2017). Non-farm economy in Madhubani, Bihar: Social dynamics and exclusionary rural transformations. *Economic and Political Weekly, 52*(25), 14–24.

Kaldor, N. (1967). *Strategic factors in economic development*. Cornell University Press.

Lanjouw, P., & Shariff, A. (2004). Rural non-farm employment in India: Access, incomes and poverty impact. *Economic and Political Weekly*, 4429–4446.

Lewis, W. A. (1954). Economic development with unlimited supplies of labor. *The Manchester School of Economic and Social Studies, 22*, 139–191.

Loison, S. A. (2015). Rural livelihood diversification in sub-Saharan Africa: A literature review. *The Journal of Development Studies, 51*(9), 1125–1138.

Losch, B., Fréguin-Gresh, S., & White, E. T. (2012). *Structural transformation and rural change revisited: Challenges for late developing countries in a globalizing world*. The World Bank.

Majid, N. (2019). Structural change and employment in India, Working Paper No. 1, ILO/SIDA partnership on employment, International Labour Organization, Geneva.

Nagler, P., & Naudé, W. (2014). *Patterns and Determinants of Non-Farm Entrepreneurship in Rural Africa: New Empirical Evidence*. Paper presented at the 9th IZA/World Bank Conference on Employment and Development in Lima, Peru.

Neves, D., & Du Toit, A. (2013). Rural livelihoods in South Africa: Complexity, vulnerability and differentiation. *Journal of Agrarian Change, 13*(1), 93–115.

OECD-FAO. (2016). Agriculture in Sub-Saharan Africa: Prospects and challenges for the next decade. OECD-FAO Agricultural Outlook 2016–2025, Rome.

OPHI. (2020). *Global Multidimensional Poverty Index 2020: Charting pathways out of multidimensional poverty: Achieving the SDGs*. Oxford: Oxford Poverty and Human Development Initiative. http://hdr.undp.org/sites/default/files/2020_mpi_report_en.pdf

Parikh, K. S., Ghosh, P. P., & Binswanger-Mkhize, H. P. (2013). Double-digit inclusive growth: Not without robust agricultural growth. *Economic and Political Weekly, 48*(51), 68–73.

Rada, C., & Von Arnim, R. (2012). Structural transformation in China and India: A note on macroeconomic policies. *Structural Change and Economic Dynamics, 23*(3), 264–275.

Rigg, J., Salamanca, A., & Thompson, E. C. (2016). The puzzle of east and Southeast Asia's persistent smallholder. *Journal of Rural Studies, 43*, 118–133.

Timmer, C. P. (2014). *Managing structural transformation: A political economy approach. WIDER Annual Lecture 18*. UNU World Institute for Development Economics Research.

UN Habitat. (2016). *Structural transformation in developing countries: Cross regional analysis*. United Nations Human Settlements Programme.

Wang, C., Yang, Y., & Zhang, Y. (2011). Economic development, rural livelihoods, and ecological restoration: Evidence from China. *Ambio, 40*(1), 78–87.

Wilson, C. (2004). Understanding the dynamics of socio-economic mobility: Tales from two Indian villages. Working Paper 236, Overseas Development Institute (ODI), London, UK.

Chapter 3
SDGs and Continued Relevance of Rural Livelihoods

Abstract With decolonization and globalization, the second half of the twentieth century saw rapid economic growth in the third world. This trend continued during the 1990 to 2015 period and saw the halving of extreme poverty and the attainment of the Millennium Development Goals (MDGs) on poverty reduction ahead of time. But this period also saw a concentration of poverty in SSA and South Asia, a rise in inequality, health, education, and living standards lagging behind economic growth, and widespread environmental degradation forcing a rethink on the global development strategy. As we move forward on the SDG goals, the real challenge for the global community is to eradicate poverty in SSA in the face of a global slowdown. While the world was already off-track to end poverty by 2030 before COVID-19, the pandemic has further caused the goal to go astray and global poverty has increased for the first time in decades. We contend that growing risks and vulnerabilities will test the sustainable attainment of SDGs even in the post-2030 era. We urge development professionals to renew their interest in rural livelihoods to tackle the present and future challenges.

Keywords MDG · SDG · Non-income goals · Anthropocene · Degradation · Natural disasters · Population · Inequality · Sustainable · Poverty depth · Resilience

3.1 The Disintegration of the Third World

During the mid-twentieth century, the countries were geopolitically divided into the first world, the second world, and the third world (Box 3.1). The first world was comprised of the USA, western Europe, Australia, and their allies. The second world comprising of the communist block, namely the Soviet Union and its East European allies. The third world comprised of all other countries not actively aligned in the cold war. However, the term 'third world' was also used to refer to the poor and developing nations of Africa, Asia, and Latin America. Over the last fifty years, the world has rapidly transformed and the narrative of development thinking has transformed and largely moved ahead of the binary concepts of first and the third world, developed and developing, rich and poor, north and south, donors and recipients as these are increasingly becoming inoperable (Mönks et al., 2017). In the twenty-first century, the third world is no longer homogenous and instead comprises a mix of low, middle, and high income countries (de Haan, 2016). Many Asian countries outperformed, escaped the third world, and joined the urbanized world. Consequently, the geography of poverty has now transformed and comprised of the three categories, namely agriculture-based, transforming, and the urbanized (Byerlee et al., 2008).

S. Tambe, *Teaching and Learning Rural Livelihoods*, Sustainable Development Goals Series,
https://doi.org/10.1007/978-3-030-90491-3_3

Box 3.1 The Three World Model

After the end of the second world war, the two superpowers—the USA and the Soviet Union were vying for global dominance. This led to geopolitical tension between the two, a period referred to as the cold war (1947–1991). The countries with close ties to the USA and the Soviet Union were largely divided into two camps or blocks—The Western Bloc: comprising of the US and its allies and The Eastern Bloc: comprising of the USSR and its allies. In 1952, the French demographer Alfred Sauvy coined the 'third world' to refer to all the countries that were not involved in the cold war and the east–west conflict. With the coining of the term 'third world', the western bloc and the eastern bloc came to be subsequently called the first and the second world, respectively. After the fall of the Soviet Union in 1991, the term 'second world' became obsolete and today, the terms 'first world' and 'third world' are used to refer loosely to the developed and the developing countries, respectively- as the 'third world' generally comprised of historically colonized and under-developed countries.

Box 3.2 The World Poverty Clock

In 2017 the World Data Lab, a Vienna based NGO developed the World Poverty Clock as a popular tool to monitor progress against poverty globally and regionally. It provides real time data across countries by using publicly available data on income distribution, production, consumption, and stratification generated by various international organizations like the UN, World Bank, and the IMF. It covers almost 99.7% of the world population and uses models to estimate poverty at the country level. The World Poverty Clock computes the speed of poverty reduction across the countries and compares it to the average speed needed to end poverty by 2030 (SDG goal-1) to categorize countries into four classes: no extreme poverty, on track, off-track and poverty is rising. In July 2021, it predicted that to meet the SDG goal of ending poverty by 2030, against the targeted poverty escape rate of 2.4 persons per second, the actual escape rate was only 0.8 persons per second.

For more information visit www.worldpoverty.io

3.2 The MDGs: Shared Global Goals

In the new millennium, the United Nations catalysed global leadership to join hands for a shared vision of eradicating poverty in its many dimensions (UNDP, 2015). This vision was translated into the Millennium Development Goals (MDGs) related to poverty, hunger, disease, schooling, gender inequality, and environmental degradation (Sachs, 2012). The MDGs were an expression of a global commitment made by the international community towards poverty reduction. The MDGs packaged these developmental priorities into an easily understandable global report card, with a set of eight goals and established measurable progress indicators (Sachs, 2012). Most of these targets were relative, with the baseline year fixed at 1990 which marked the beginning of the global preparatory effort. The MDGs expressed the desired ends, i.e. a world with less hunger, less illiteracy, less disease, and an end to gender discrimination Vandemoortele (2009). They did not however prescribe a particular means to achieve these goals as that would be infringing upon the sovereign domain of the member states. These goals steered developmental priorities across the world and helped promote global awareness, political ambition, developmental action, and public pressure (Sachs, 2012). For close to two decades, the MDGs remained the focus of global policy debates, national policy planning, and guided the work of development organizations and civil society (Sachs, 2012). Thus the MDGs made

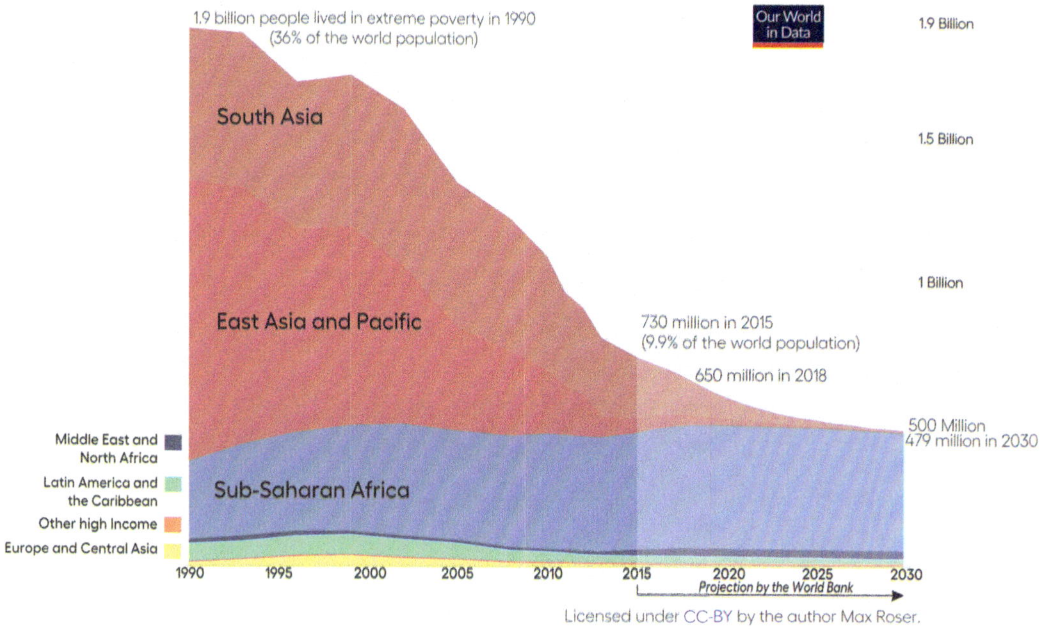

Fig. 3.1 Decline in global poverty region wise with future projections (Source: Our World in Data)

a lasting contribution in forging global policy cooperation and bringing together member countries with the shared vision of fighting poverty in all its dimensions and improving the quality of lives of the poor. In the subsequent subsections, we discuss the impact of MDGs on global poverty, the emerging trends, and the unmet challenges leading to a course alteration in the form of the Sustainable Development Goals (SDGs).

3.2.1 *Making a Dent on Global Poverty*

This period from 1990 to 2015 saw the most rapid decline in global poverty in history and the Millennium Development Goals (MDGs) on poverty reduction were achieved ahead of time (Fig. 3.1) (UNDP, 2015). Between 1990 and 2015, the number of extreme poor, i.e. those living on less than $1.25 per day declined from 1.9 billion to 836 million (UNDP, 2015). In these 25 years, more than one billion people escaped extreme poverty as the global poverty rate fell from 36% to 11% (UNDP, 2015). This period also saw the emergence of a global middle class thanks to the rapid expansion of the economy in developing countries. The decline was led by Asia, with China and India being the front runners. This poverty eradication had two main drivers—rapid and sustained economic growth and redistributive programmes (Page & Pande, 2018). It was driven by a rise in non-farm livelihoods across Asia and Africa due to opportunities created by structural adjustment and market liberalization policies and resulted in widespread rural income diversification away from agriculture pursuits (Bryceson, 2002; IDFC Rural Development Network, 2013). This achievement was contested as misleading by some as the baseline of the analysis was shifted back to 1990 instead of 2000 when the MDGs began. This resulted in the MDGs claiming the achievements of the 1990–2000 decade when China made rapid strides in poverty alleviation which could not be attributed to it (Hickel, 2016).

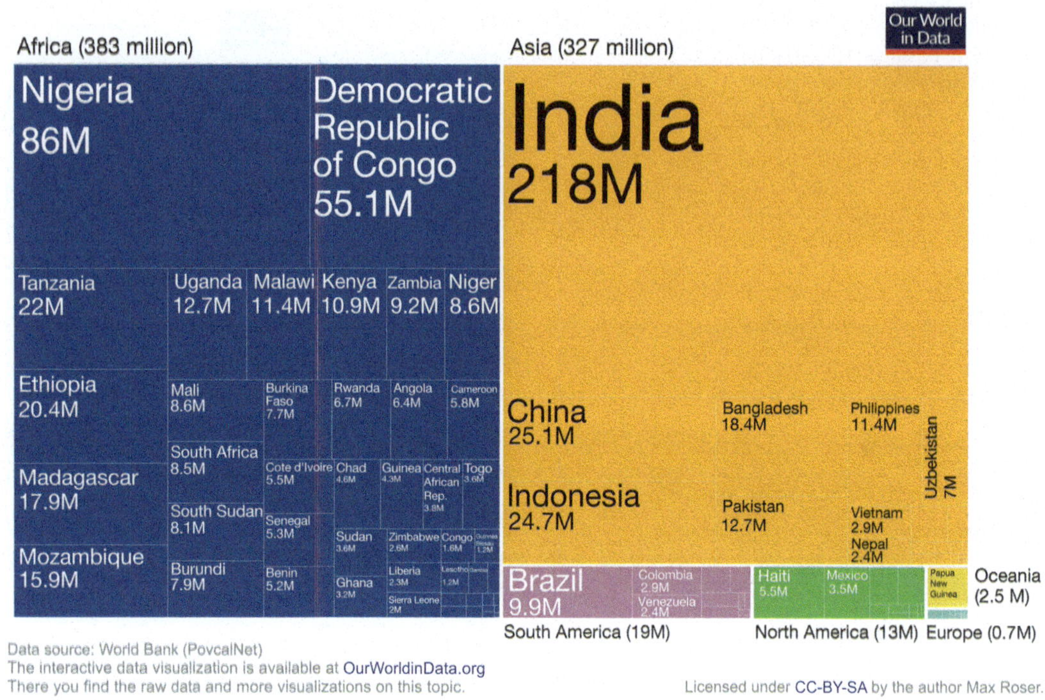

Fig. 3.2 Global poverty distribution across regions and countries in 2013

3.2.2 Poverty Concentrating in SSA and South Asia

While impressive gains were made in reducing global poverty, it remains embedded particularly in SSA and South Asia (Fig. 3.2). These two regions along with East Asia and the Pacific account for nearly 95% of global poverty (Cruz et al., 2015). South Asia achieved rapid poverty reduction during this period but SSA lagged, resulting in a rise in its share of global poverty to 43% (Cruz et al., 2015). While the poverty rate in the low-income countries is higher, surprisingly it is in the lower-middle-income countries such as India, Indonesia, and Nigeria where most of the global poor reside. When accounting for poverty depth, the geographical concentration of global poverty has further shifted to SSA with the region accounting for 57% share (Cruz et al., 2015). There is a need to now shift the policy discourse towards the plight of the ultra-poor and their special needs in these regions that are trailing.

3.2.3 Lagging in Non-Income Goals

Though significant progress on income poverty was made during the MDG period, the goals related to the non-income dimensions such as infant mortality rate, maternal mortality rate, access to drinking water, sanitation facilities, etc., saw unequal progress between and within regions (Fig. 3.3 and 3.4) (Cruz et al., 2015). Gender inequality persists with only half of the working-age women able to participate in the labour force, earn 24% less than men and have much lower representation in decision making (UNDP, 2015). The poor in developing regions still face disproportionate challenges in accessing nutrition, educational services, health facilities, and basic amenities (UNDP, 2015).

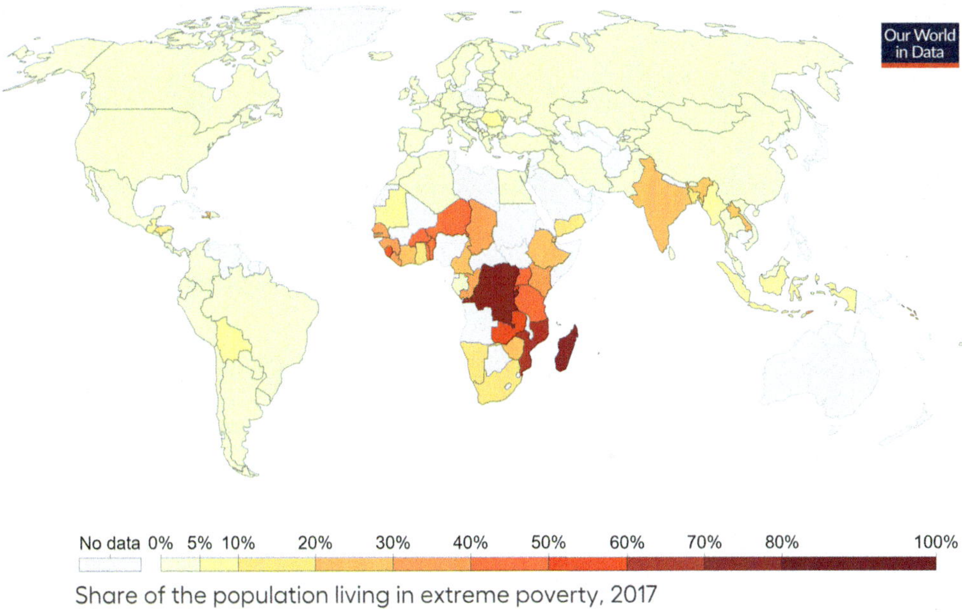

Fig. 3.3 Share of population living in extreme poverty in 2017

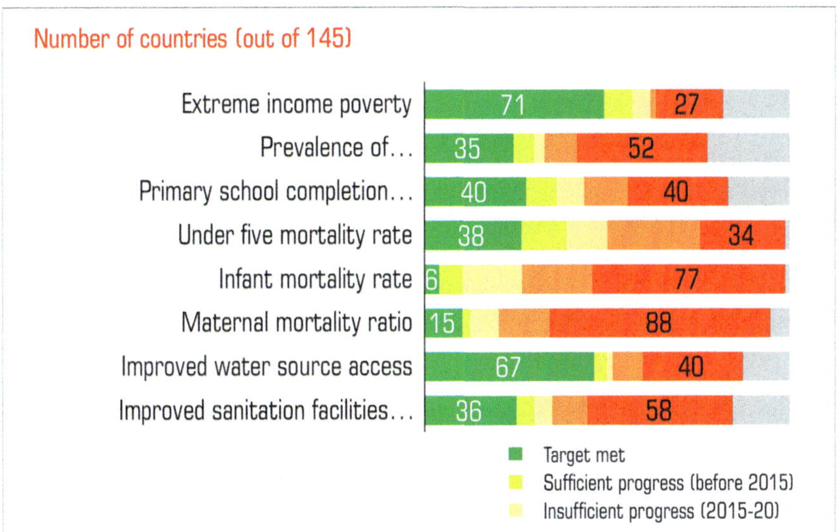

Fig. 3.4 Unequal progress on non-income goals between 1990 and 2015 (Source: Cruz et al., 2015)

3.2.4 Severe Decline in Natural Capital

The MDG period which heralded an unprecedented upswing in human wellbeing also witnessed a severe decline in natural capital. The rise in global prosperity saw the natural capital of the earth severely eroded and precipitated the dawn of a new geological epoch—the Anthropocene. The single-minded focus only on anthropocentric gains came at a cost, as planetary boundaries were breached, ecosystem services were compromised, environment pollution peaked, climate change accelerated,

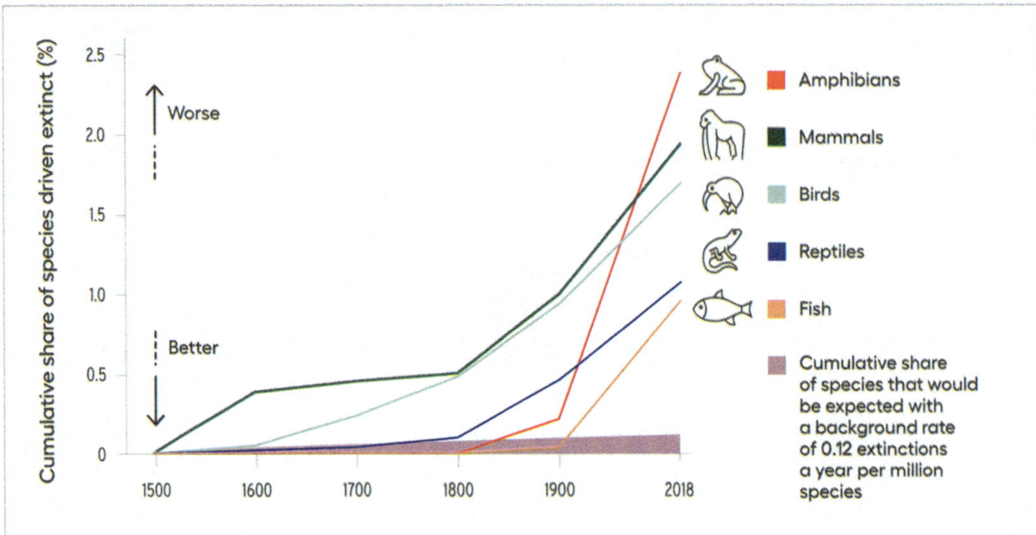

Fig. 3.5 Over the last few centuries, the rate of species extinction has accelerated and is estimated to be several times more than the background rates (Source UNDP HDR, 2020)

coupled with biodiversity loss, and conversion of forests (Sachs, 2012). The earth's support systems including the oceans, forests, atmosphere, waterways, biodiversity, and biogeochemical cycles that are vital for human survival were severely impacted (Griggs et al., 2013). There was also a realization that further disruption of the earth's support systems would result in widespread humanitarian crises due to shortages in water, food, energy, and crippling natural disasters.

As three-quarters of the land-based environment and about two-third of the marine environment have been significantly altered by human actions, nature is declining globally at rates unprecedented in human history—and the rate of species extinctions is accelerating, with grave impacts on people around the world (Fig. 3.5) (IPBES, 2019; UNDP HDR, 2020). Rockström et al. (2009) developed a planetary boundaries concept involving earth system processes to develop a safe operating space for humanity (Fig. 3.6). They identified nine planetary boundaries and found that humanity had already transgressed three planetary boundaries for climate change, the rate of biodiversity loss, and the changes to the global nitrogen cycle. The ecological footprint metric measures how much nature we have against how much we use. On this front as well, we use the equivalent of 1.6 earths to provide the resources we use and absorb our waste (Wackernagel & Rees, 1998). The Global Futures report calculates the economic costs of inaction on the climate and ecological crisis front by assessing the value of just six ecosystem services as USD 9.87 trillion in GDP by 2050 with the decline being felt unevenly across the globe (Fig. 3.7) (Johnson et al., 2020).

3.2.5 *Rising World Population and Inequity*

Beyond the environmental threats and declining natural capital, humanity faces another serious threat in the post-MDG era, in the form of the increasing world population (Fig. 3.8). The human population continues to grow rapidly, by about 75–80 million people per year and is on a trajectory to reach 9 billion by the middle of the twenty-first century, and even touch 10–11 billion by the end of the century (Sachs, 2012; UN DESA, 2019). Rising population along with the rapidly increasing incomes per person in large emerging economies like India and China will lead to an increase in the demand for

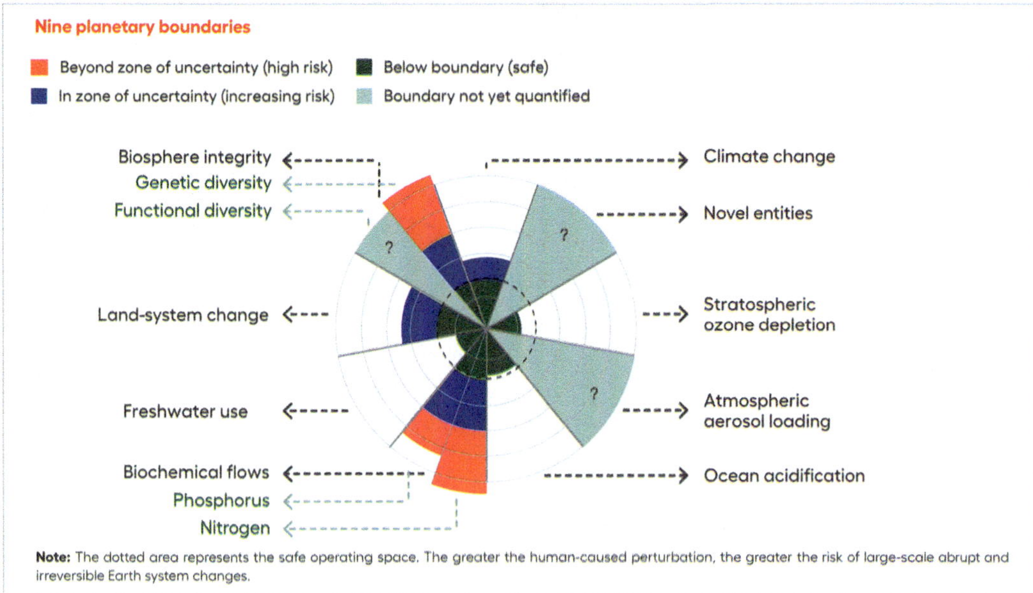

Fig. 3.6 The safe operational space for humans (Source: UNDP HDR, 2020)

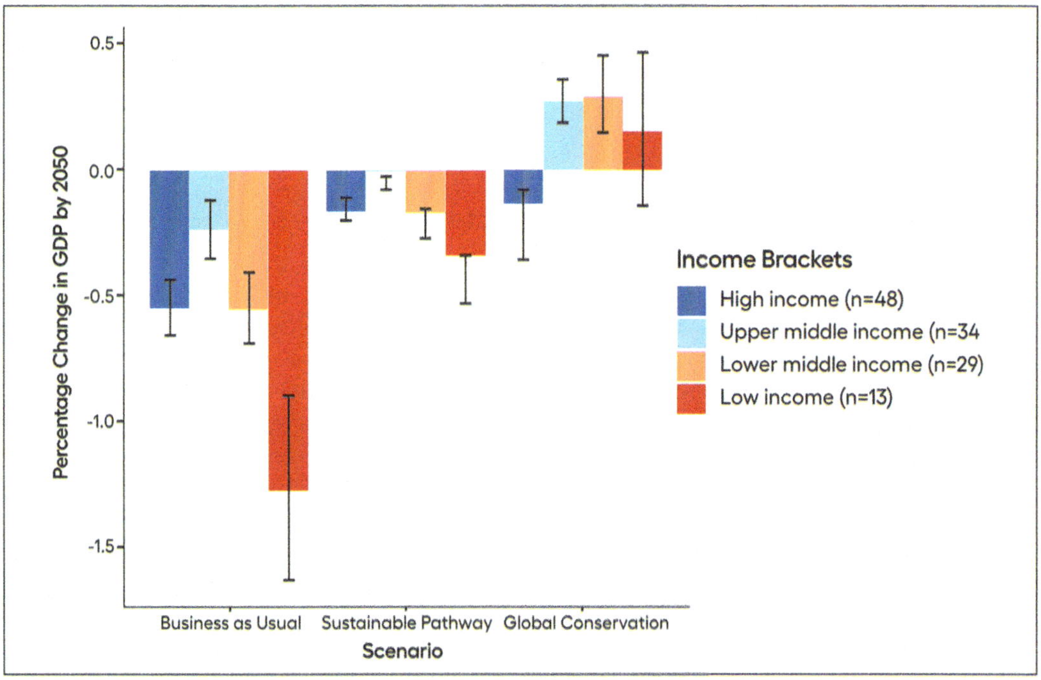

Fig. 3.7 Percentage change in GDP due to changes in all ecosystem services for three 2050 scenarios (Source: Johnson et al., 2020)

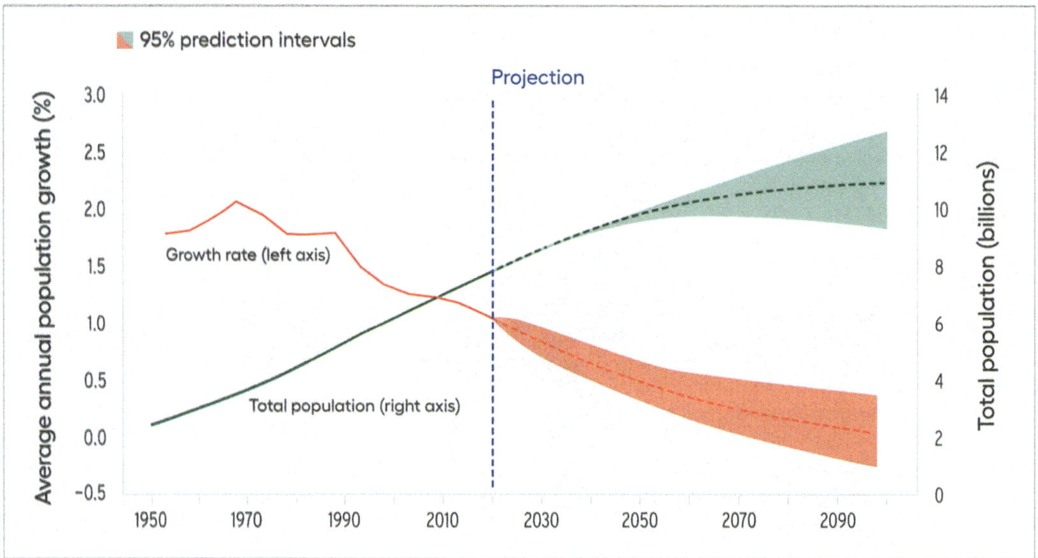

Fig. 3.8 The world population is growing, but at a slower rate (Source UNDP HDR 2020)

food and feed grains. It is already being witnessed that the rapid gains in farm productivity attained during the green revolution period have slowed worldwide. Increasing grain production threatens to destruct natural habitats, cause climate change, water stress, increased fertilizer pollution, decreased biodiversity, and more. The social outcomes of this could be deeply destabilizing, threatening to push millions into chronic hunger (Sachs, 2012).

3.3 SDGs: Charting a Sustainable Trajectory

While MDGs made rapid gains in combating global poverty, they had many shortfalls as well. This realization was in the minds of world leaders when they met at the United Nations Rio + 20 summit in 2012 to design the Sustainable Development Goals (SDGs) as a follow up to the MDGs. The SDGs would need to work not only for people but also for the planet by charting out a sustainable trajectory. So the challenge for the SDGs was to ensure the triple bottom line of economic prosperity, social inclusion, and environmental sustainability (Sachs, 2012). The SDGs lay out a set of goals to achieve a sustainable future for all by 2030. They focus on 17 global goals highlighting time-bound targets for prosperity, people, planet, peace, and partnership (Sachs et al., 2019). The SDGs are different from MDGs in the sense that they put sustainability at the centre of the developmental agenda and emphasize the interconnected environmental, social, and economic aspects of development (Schleicher et al., 2018). They are a result of the growing recognition of the relevance of the natural environment for human wellbeing (Schleicher et al., 2018). While the MDGs focussed mostly on social and economic goals, the SDGs attempt to integrate environment and development (Elder & Olsen, 2019). The MDGs were targets mainly for poor countries to which the rich countries were to add their solidarity and assistance through the transfer of aid and technology. The SDGs on the contrary are more comprehensive with a broad spectrum and targets and responsibilities for all countries rich or poor (Sachs, 2012).

The SDGs laid out ambitious targets of ending poverty and hunger in all forms everywhere by 2020. This aspirational goal faces new types of challenges that will need to be addressed to make the

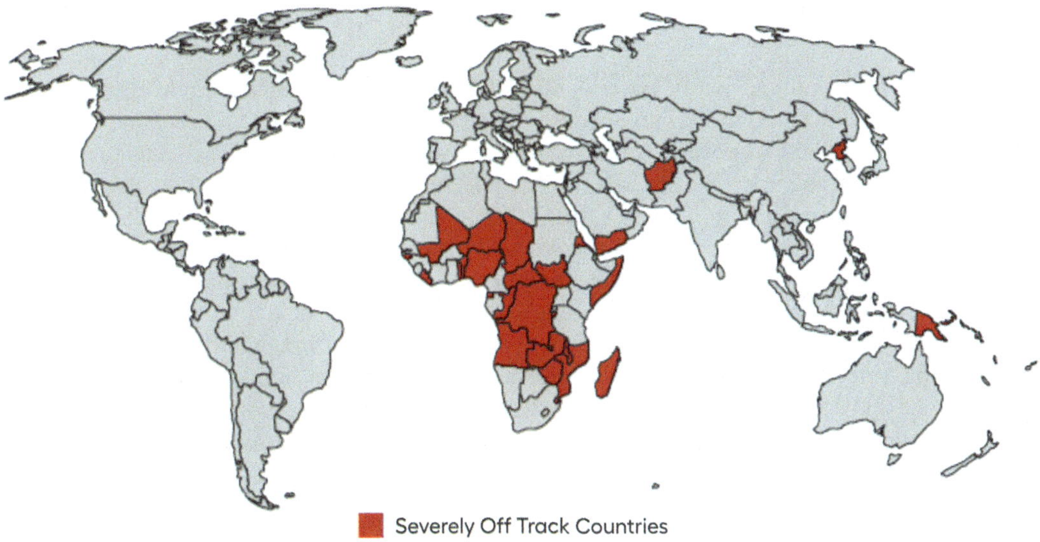

Fig. 3.9 Global map of the Severely Off-Track Countries (SOTC) countries (Source: Gertz & Kharas, 2018)

world poverty-free. Despite halving extreme poverty, the progress across countries has been uneven, with several smaller countries mostly in sub-Saharan Africa witnessing stagnation. Many of these fragile states are ridden with conflicts and remain at the bottom of the development ladder. Around 800 million people still languish in poverty concentrated in sub-Saharan Africa and South Asia. While some of them are in countries that have low levels of poverty and have policies and programmes in place to eradicate poverty by 2030, many reside in countries that face multiple obstacles such as low incomes, sluggish economic growth, fragile governance, conflicts, and a poverty rate of over 20% (Chandy, 2017; Marcus et al., 2018). 'Leaving no one behind' is the recurring and overarching objective of the 2030 global agenda. While during the MDGs period, poverty concentrated in large, growing Asian economies was reduced, the SDG period is confronted with entrenched poverty in smaller African countries facing governance challenges. Gertz and Kharas (2018) identified 31 severely off-track countries (SOTC) mostly from Africa and identify the four underlying obstacles to development that contribute to persistent poverty, namely low government effectiveness, weak private sector, conflict and violence, natural hazards, and environmental risks (Fig. 3.9). Page and Pande (2018) argue that economic growth and aid will not be sufficient to end global poverty by 2030 as it persists in countries where governance is weak and steps are first needed to strengthen the capacity of these states. On the environment front, the real challenge for the SDGs is to demonstrate positive change on-scale driven by climate action and biodiversity action.

3.4 SDGs: End Poverty in all Forms Everywhere

The first goal of the SDGs is to end global poverty in all its forms everywhere. This goal comprises of six targets, namely—'*eradicating extreme poverty, reduce by half the incidence of multidimensional poverty, ensure substantial coverage of the poor under social protection systems, ensure that the poor and vulnerable have equal rights to economic resources and basic services, build the resilience of the poor and reduce their exposure and vulnerability to shocks and disasters, implement programmes and policies to end poverty in all its dimensions, and support accelerated investments in poverty eradication actions*'.

3.5 Challenges in Ending Global Poverty

While MDGs lifted millions out of poverty, SDGs have to lift the remaining while also mitigating their vulnerability of falling back into poverty. The SDG period will have to overcome challenges related to a greater depth of poverty of the remaining poor, the multidimensional nature of poverty, dealing with fragile countries, the setback caused by COVID-19, and growing vulnerabilities due to disasters and diseases. During the MDG period, while significant progress was made in reducing poverty in several parts of the world, a lot of work still needs to be done. Poverty elimination in the SDG period needs to also account for the depth of poverty of the remaining poor (Cruz et al., 2015). The poor households in the SDG period are deeper in poverty, especially in sub-Saharan Africa. Hence, poverty alleviation efforts around the globe will face diminishing returns as those still poor have multiple vulnerabilities. The patterns of global poverty are also changing, it is now concentrated in south Asia and SSA. Most of the poor now live in middle income countries with the poorest in SSA. While poverty in the middle income countries has declined, there is persisting inequality.

Poverty is complex and cannot be measured in terms of incomes alone, and hence a multidimensional poverty index (MPI) was developed by the Oxford Poverty and Human Development Initiative (OPHI) and the United Nations Development Programme (UNDP). The global MPI measures the deprivations that a person faces concerning education, health, and living standards. Hence, it complements the traditional income-based poverty measures and is strongly related to the SDG goals. The 2020 global MPI report estimates that there are a total of 1.3 billion (22%) MPI poor people of which 84.3% live in South Asia and sub-Saharan Africa (SSA). Also, nearly 46% of these multidimensionally poor live in severe poverty (OPHI, 2020).

3.6 Future Projections of Global Poverty

All regions are on track for ending extreme poverty by 2030, but for SSA where the number of poor is increasing. In 2016, when the SDG era started, Africa accounted for just over 60% of global poverty. Today, it is over 75% and by 2030; it could be close to 90% (Kharas et al., 2018). As we move

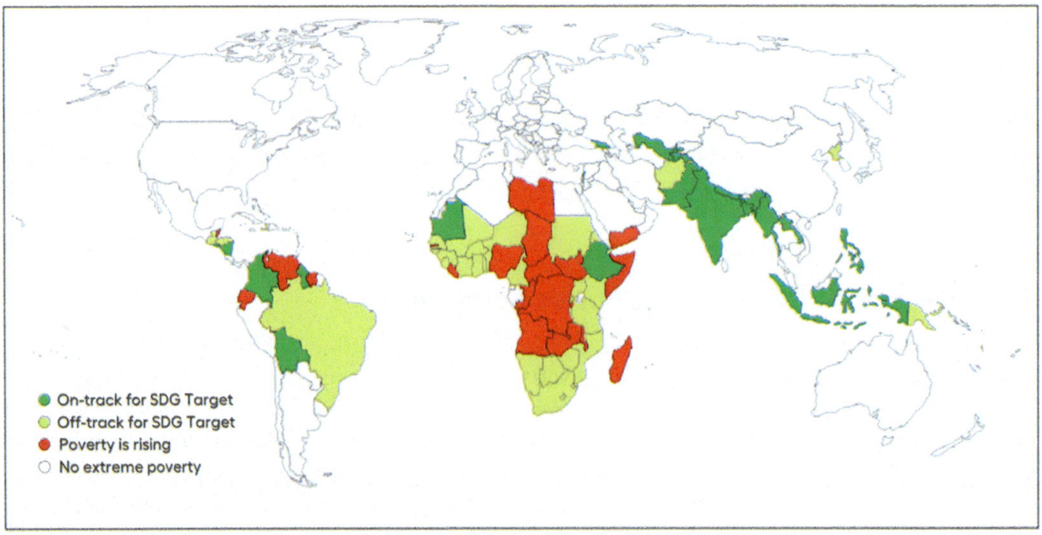

Fig. 3.10 Future poverty eradication status of countries (Source: Cuaresma et al., 2018)

forward to end extreme poverty, Africa will remain the last frontier in this global effort (Kharas et al., 2018). The slowdown in the global economy has negatively impacted the pace of poverty reduction which has come down from 1 person per second in 2017 to 0.6 person per second in 2019 (Kharas et al., 2018). Cuaresma et al. (2018) classified 24 countries with 207 million poor people as on track to become poverty-free before 2030. However, they identified 40 off-track countries where 131 million would remain poor by 2030. And even worse, in 20 countries with 242 million poor people, they estimate that absolute poverty will rise (Fig. 3.10).

3.7 Impact of Covid-19 Pandemic on Global Poverty

The pre-COVID-19 projection suggested that 6% of the global population would still be living in extreme poverty in 2030, falling short of the target of poverty elimination (UN DESA, 2020). The COVID-19 pandemic resulted in a serious setback to the progress towards SDGs as it has resulted in a contraction of economies with most countries reporting negative growth. Political leaders are confronted with the unenviable choice to decide whether to continue with the lockdown restrictions or to open up the economy. Decisions will result in either loss of lives or livelihoods. Sumner et al. (2020) estimate that due to COVID-19, contractions in household incomes could result in global poverty increasing for the first time in several decades with some regions sliding back to the poverty levels of the 1990s. The UN DESA (2020) projects that in the worst case scenario the COVID-19 could push 71 million people into poverty. The increase in deprivations because of COVID-19 would set multidimensional poverty reduction back by 5–9 years, with an additional 237–490 million people falling into multidimensional poverty as the MPI increases from the present 0.095 to between 0.125 and 0.156 across 70 countries (OPHI, 2020). The head of the World Food Program, the winner of the 2020 Nobel Peace Prize, warned that due to the COVID-19 pandemic humanitarian crisis is worsening, and consequently, the number of people marching toward starvation has jumped from 135 million to 270 million. Growing risk and vulnerability due to natural disasters, pandemics, environmental change, and other unpredictable events will increasingly test the sustainability of this attainment.

3.8 The Way Forward

The COVID-19 pandemic has resulted in wiping away some of the gains made over the last decade in achieving development targets. Consequently, to account for the new normal, there have been calls to recalibrate the SDG targets. Naidoo and Fisher (2020) point out that COVID-19 has shifted the very foundations on which the attainment of the SDG goals relied on—sustained economic growth and global developmental aid. The global economy is expected to contract and poverty levels in the global south are set to rise for the first time in several decades. This coupled with a drop in overseas developmental aid will make the attainment of two-third of the SDG goals unlikely (Naidoo & Fisher, 2020). Sachs et al. (2020) have argued that it is not a time to lower our ambition and weaken or abandon the time-bound SDGs. Instead, we need to redouble our efforts in favour of human rights and environmental sustainability. The momentum generated during the MDG period needs to be sustained. The next few decades are critical and will determine whether our successes are sustainable or will get undermined or reversed (Fankhauser & Stern, 2016). Sustaining the gains achieved during the first few decades of this millennium will continue to keep rural livelihoods relevant for several more decades if not more. Also, despite the rapid urbanization especially in the global south, it is estimated that in 2050 about 2.8 billion people will continue to reside in rural areas having dependence on agriculture with Asia and SSA comprising two-third of this global rural population (Losch et al., 2012).

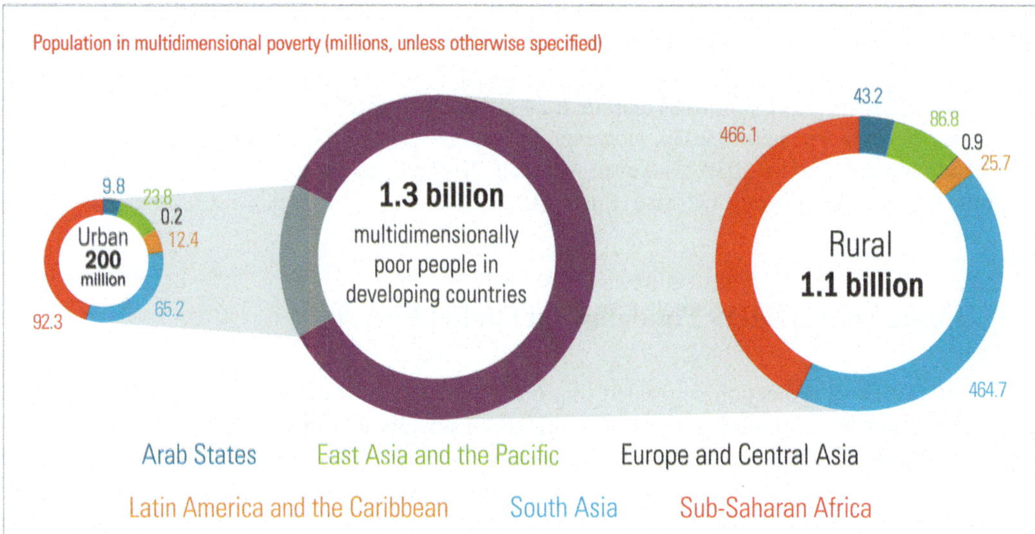

Fig. 3.11 Of the total 1.3 billion people who are multidimensionally poor, 84.2 percent live in rural areas (Source: OPHI, 2020)

Extreme poverty is predominately a rural phenomenon, with two-thirds of the extreme poor employed in agriculture and extreme poverty rates are four times higher in rural areas compared to urban areas (ILO, 2016). In every development region of the world, multidimensional poverty is concentrated in rural areas (OPHI, 2020). Of the total 1.3 billion who are multidimensionally poor, 84.2% live in rural areas, equally shared between South Asia and SSA (Fig. 3.11) (OPHI, 2020). However, the rural economy holds considerable potential for economic growth, employment creation, and promotion of decent work if the right policies are in place. Thus, rural areas in South Asia and SSA continue to remain the final frontier in the achievement of the SDG Goal-1 of ending poverty in all forms everywhere.

While the SDGs have renewed the focus on ending poverty and hunger, there is a waning interest in rural livelihoods (Scoones, 2009, 2015). Contribution of livelihood approaches in cross-cutting areas such as climate change adaptation, vulnerability assessment, and in emerging areas such as resilience approaches, urban studies, etc., especially in the global South will help in keeping it relevant in these times of urbanization, global environmental change, and sustainable development. Even though the livelihoods idea is central to the goal of ending poverty and hunger, there is a need to refocus and re-engage with this approach prominently in academics, policymaking, and development.

Exercises

1. Analyse the global progress made on the sustainable development front during the MDG period using the data available at https://ourworldindata.org/? Critically analyse which countries showed good progress, and which lagged and why?
2. Using the data available at https://ourworldindata.org/, analyse which is a better indicator to measure environmental progress—carbon emissions per capita or material footprint per capita? Also, discuss whether the development trajectory of the high income countries follows the Environment Kuznets Curve?
3. Draw a radar graph depicting the social, economic, and environmental progress of the low, medium, and high income nations during the MDG period (1990 and 2015) using the data available at https://ourworldindata.org/ and in other sites. Which indicators would you select to represent each of these three dimensions and why?

4. Using the radar graphs from the previous question, analyse the development trajectory of the high-, medium-, and low-income countries during the MDG period (1990 and 2015).
5. Analyse the global progress made on the sustainable development front during the initial years of the SDG period using the data available at https://ourworldindata.org/? Which countries showed good progress, and which lagged?
6. Is high human development directly correlated to a high per capita material footprint? Draw the graph of material footprint per capita versus HDI using the data available at https://ourworldindata.org/ and http://hdr.undp.org/en/2020-report to identify countries that show high human development with relatively lower material footprint and analyse the reasons and vice versa.
7. The Human Development Index (HDI) was constructed by Mahbub Al Haq at UNDP in 1990 to take into account certain aspects of quality of life. Compare the GDP per capita with HDI for select countries and discuss why HDI is a better indicator.
   http://hdr.undp.org/en/2020-report
8. Extreme poverty is defined by the 'international poverty line' as living on less than $1.90 per day by the world bank, while the multidimensional poverty index (MPI) is a composite indicator constructed from ten indicators representing the three dimensions of health, education, and living standards. Compare the extreme poverty map and the multidimensional poverty map published by the Oxford Poverty & Human Development Initiative (OPHI) using the data available at https://ourworldindata.org/ and http://hdr.undp.org/en/2020-MPI. Analyse the similarities and differences for select countries across the income gradient in these two maps and discuss the reasons behind the same.
9. The Human Development Report (HDR) 2020 in Fig. 3.8 presents an adjustment to the HDI for planetary pressures. This planetary pressures adjusted HDI (PHDI) corrects the HDI by taking into account a country's per capital level of carbon dioxide emissions and material footprint. Compare the inequalities adjusted HDI and the PHDI for select countries across the income gradient and discuss the reasons behind the same.
   http://hdr.undp.org/en/2020-report
10. The Human Development Report (HDR) 2020 in Fig. 3.5 presents two simulations of the Shared Socio-economic Pathways (SSP) scenarios. The SSP5 or the business-as-usual scenario would move all the five world regions to high-income status but with global warming of 3–5 degrees Celsius above preindustrial levels. On the other hand, the SSP1 scenario would bring all the regions to a space that combines high levels of income with global warming within 2 degrees Celsius above preindustrial levels. Discuss the steps needed by the developing and developed world to decouple from SSP5 and align with SSP1.
    http://hdr.undp.org/en/2020-report
11. In today's modern world which is rapidly urbanizing, is the theme of rural livelihoods still relevant? Give reasons to support your answer.

Additional resources

- IFAD and the SDGs https://www.ifad.org/en/ifad-and-the-sdgs

References

Bryceson, D. F. (2002). The scramble in Africa: Reorienting rural livelihoods. *World Development, 30*(5), 725–739.

Byerlee, D., De Janvry, A., Sadoulet, E., Townsend, R., & Klytchnikova, I. (2008). *World development report 2008: Agriculture for development (English). World development report; no. 30.* World Bank Group.

Chandy, L. (2017). *No country left behind: The case for focusing greater attention on the world's poorest countries.* The Brookings Institution.

Cruz, M., Foster, J., Quillin, B., & Schellekens, P. (2015). *Ending extreme poverty and sharing prosperity: Progress and policies. Policy Research Note PRN/15/03 (2015).* World Bank Group.

Cuaresma, J. C., Fengler, W., Kharas, H., Bekhtiar, K., Brottrager, M., & Hofer, M. (2018). Will the sustainable development goals be fulfilled? Assessing present and future global poverty. *Palgrave Communications, 4*(1), 1–8.

de Haan, L. (2016). Livelihoods in development. *Canadian Journal of Development Studies/Revue canadienned'études du développement, 38*(1), 22–38.

Elder, M., & Olsen, S. H. (2019). The design of environmental priorities in the SDGs. *Global Policy, 10*, 70–82.

Fankhauser, S., & Stern, N. (2016). Climate change, development, poverty and economics. *The State of Economics, the State of the World.*

Gertz, G., & Kharas, H. (2018). *Leave no country behind: Ending poverty in the toughest places. Global Economy & Development Working Paper 110.* Brookings Institution.

Griggs, D., Stafford-Smith, M., Gaffney, O., Rockström, J., Öhman, M. C., Shyamsundar, P., … Noble, I. (2013). Sustainable development goals for people and planet. *Nature, 495*(7441), 305–307.

Hickel, J. (2016). The true extent of global poverty and hunger: Questioning the good news narrative of the millennium development goals. *Third World Quarterly, 37*(5), 749–767.

IDFC Rural Development Network. (2013). *India rural development report 2012–13.* Orient BlackSwan.

ILO. (2016). *Transforming jobs to end poverty.* International Labour Organization, World Employment Social Outlook (WESO).

IPBES. (2019). In E. S. Brondizio, J. Settele, S. Díaz, & H. T. Ngo (Eds.), *Global assessment report on biodiversity and ecosystem services of the intergovernmental science-policy platform on biodiversity and ecosystem services.* IPBES secretariat.

Johnson, J., Baldos, U. L., Hertel, T., Nootenboom, C., Polasky, S., & Roxburgh, T. (2020). Global Futures: Modelling the global economic impacts of environmental change to support policy-making-technical report, (WWF UK). Purdue University, West Lafayette, IN: Global Trade Analysis Project (GTAP). Retrieved on 31st July, 2021 at 21:00 from https://www.gtap.agecon.purdue.edu/resources/res_display.asp?RecordID=6186

Kharas, H., Hamel, K., & Hofer, M. (2018). Rethinking global poverty reduction in 2019. *Blog. Future Development at Brookings, 13.*

Losch, B., Fréguin-Gresh, S., & White, E. T. (2012). *Structural transformation and rural change revisited: Challenges for late developing countries in a globalizing world.* The World Bank.

Marcus, M., Desai, H., Samman, E., & Evans, M. (2018). *Financing the end of extreme poverty.* Overseas Development Institute Report. https://www.odi.org/sites/odi.org.uk/files/resource-documents/12411.pdf

Mönks, J., Carbonnier, G., Mellet, A., & De Haan, L. (2017). Towards a renewed vision of development studies. *International Development Policy| Revue internationale de politique de développement*, 8(8.1).

Naidoo, R., & Fisher, B. (2020). Reset sustainable development goals for a pandemic world. *Nature, 583*(7815), 198–201.

OPHI. (2020). *Global multidimensional poverty index 2020: Charting pathways out of multidimensional poverty: Achieving the SDGs.* Oxford Poverty and Human Development Initiative. http://hdr.undp.org/sites/default/files/2020_mpi_report_en.pdf

Page, L., & Pande, R. (2018). Ending global poverty: Why money isn't enough. *Journal of Economic Perspectives, 32*(4), 173–200. https://pubs.aeaweb.org/doi/pdf/10.1257/jep.32.4.173

Rockström, J., Steffen, W., Noone, K., Persson, Å., Chapin, F. S., III, Lambin, E., … Nykvist, B. (2009). Planetary boundaries: Exploring the safe operating space for humanity. *Ecology and Society, 14*, 2.

Sachs, J. D. (2012). From millennium development goals to sustainable development goals. *The Lancet, 379*(9832), 2206–2211.

Sachs, J. D., Schmidt-Traub, G., Mazzucato, M., Messner, D., Nakicenovic, N., & Rockström, J. (2019). Six transformations to achieve the sustainable development goals. *Nature Sustainability, 2*(9), 805–814.

Sachs, J., Schmidt-Traub, G., & Lafortune, G. (2020). Speaking truth to power about the SDGs. *Nature, 584*(7821), 344.

Scoones, I. (2009). Livelihoods perspectives and rural development. *The Journal of Peasant Studies, 36*(1), 171–196. https://doi.org/10.1080/03066150902820503

Scoones, I. (2015). *Sustainable livelihoods and rural development.* Practical Action Publishing.

Schleicher, J., Schaafsma, M., & Vira, B. (2018). Will the sustainable development goals address the links between poverty and the natural environment? *Current Opinion in Environmental Sustainability, 34*, 43–47.

Sumner, A., Hoy, C., & Ortiz-Juarez, E. (2020). *Estimates of the impact of COVID-19 on Global Poverty* (pp. 800–809). UNU-WIDER.

UNDP. (2015). *The millennium development goals report 2015.* United Nations. ISBN 978-92-1-101320-7.

UNDP HDR. (2020). *The next frontier, human development and the anthropocene, human development report 2020.* UNDP.

UN DESA. (2019). *World population prospects 2019, United Nations Department of Economic and Social Affairs Population Division, Highlights (ST/ESA/SER.A/423).* Department of Economic and Social Affairs, United Nations.

UN DESA. (2020). *Sustainable development outlook 2020, achieving SDGs in the wake of COVID-19: Scenarios for policymakers.* Department of Economic and Social Affairs, United Nations.

Vandemoortele, J. (2009). The MDG conundrum: Meeting the targets without missing the point. *Development policy review, 27*(4), 355–371.

Wackernagel, M., & Rees, W. E. (1998). *Our ecological footprint: Reducing human impact on the earth.* New Society Publishers.

Part II
Synthesis of the Present State of Knowledge

This part forms the core of the book where we describe the analytical frameworks of the various rural livelihood approaches developed over the last five decades. We devote six chapters to assimilate the six main rural livelihood approaches, and synthesize the learnings in the last chapter. We describe the sustainable livelihoods approach, common pool resources and livelihoods, livelihood trajectories, rights-based approach, graduation approach for the ultra-poor, and the resilience framework to address complex risks. For each of these analytical frameworks, we discuss the conceptual part, its constituent elements, the process it prescribes, the insights it unravels, its limitations, and applicability in science, policy, and practice. This section is especially designed keeping in mind the requirements of educators and students in teaching and learning rural livelihoods. At the end of each chapter, we offer a set of exercises to bring about a deeper understanding of these frameworks and their applications in real-life situations. The students can apply a single livelihood approach or explore using a combination of approaches while crafting their answers. In the last chapter, we compare these frameworks and discuss how they supplement, complement, and contrast with each other, and the possibilities of fusing them.

Chapter 4
Sustainable Livelihoods Approach

Abstract The sustainable livelihood approach is a way of thinking that rests on a set of core principles and provides a conceptual framework for analysis. The sustainable livelihoods thinking emerged in the 1990s aligned with the emerging development approaches of participatory, bottom-up planning that challenged the dominant economic growth-led development paradigm. It recognizes that the poor have resources, capabilities, and the agency, and draw on the concepts of the capability approach, equity, and environmental sustainability. This approach was widely adopted in international development policy and practice and resulted in a shift from a top-down sectoral approach to household-level investments. Over time, this approach was criticized for being too household focussed and abstract to provide insights on general patterns and policy analysis. That it does not automatically lend itself to macro-level analysis, comparative studies, impact evaluation, and policy analysis. However, over time several studies have innovated to combine this framework with participatory methods and quantitative techniques to enhance its utility in livelihood assessment, project management, and impact evaluation.

Keywords Economic growth · Wellbeing · Vulnerability · Livelihood assets · Policy analysis · Livelihood strategy · Livelihood dynamics · Livelihood assessment · Project planning · Monitoring · Evaluation

4.1 Sustainable Livelihoods Approach

During the 1980s and 1990s, there was a shift in the development approaches from economic growth and structural reforms to human wellbeing and sustainability as propounded in the Brundtland Commission Report in 1987 and the first UNDP Human Development Report in 1990 (Solesbury, 2003). The sustainable livelihoods approach (SLA) emerged in the 1990s in response to growing disillusionment with the neoliberal development agenda and the accompanying macro-economic, top-down, interventionist responses to development which looked at people as passive recipients (Donohue & Biggs, 2015). The historical origins of SLA can be traced to the seminal work of Nobel laureate Amartya Sen's seminal work on capabilities and wellbeing. He described *functionings* as the various activities that an individual takes up from amongst a larger set of *capabilities* one possesses, and the *freedom* to choose a life one would like to lead (Sen, 1982, 1985, 1993). This freedom is however not absolute and is dependent on social arrangements and the larger policy environment provisioned by the state. The SLA was well aligned with the global shift in worldview and succeeded in catching the attention of donor agencies with its people-centred, bottom-up approaches and became the mainstream international development approach in the 1990s (Small, 2007).

S. Tambe, *Teaching and Learning Rural Livelihoods*, Sustainable Development Goals Series,
https://doi.org/10.1007/978-3-030-90491-3_4

The concept of sustainable livelihoods was first articulated by Robert Chambers of the Institute of Development Studies (IDS) and subsequently adopted in international development as:

'A livelihood comprises the capabilities, assets (including both material and social resources), and activities required for a means of living. A livelihood is sustainable when it can cope with and recover from stresses and shocks and maintain or enhance its capabilities and assets both now and in the future, while not undermining the natural resource base' (Chambers & Conway, 1992; Ashley & Carney, 1999; DFID, 1999). The essence of this definition is that livelihoods are more than income generation, and the need to acknowledge the agency of the poor as they too have resources and capabilities to draw upon (Small, 2007). This approach integrated the capabilities approach with the concepts of equity and sustainability and challenged the dominant development paradigm of a welfare state that focussed on production, employment, and poverty (Solesbury, 2003). It signified a shift away from sectoral and structural interventions to focussing on the assets and capabilities of the poor (Carney, 1998; Scoones, 1998). It in a way advocated that while economic growth was necessary for poverty reduction, but the poor also require capabilities to take advantage of this growth (Donohue & Biggs, 2015). Several international agencies such as DFID, UNDP, Oxfam, CARE, and others adapted this conceptualization as per their institutional priorities (Carney, 2003).

4.2 Sustainable Livelihood Principles

The SLA rests on the core principles that stress people-centred, responsive, and multi-level approaches to development to enhance the progress in poverty reduction (Ashley & Carney, 1999). These principles listed in Box 4.1 capture the essence of sustainable livelihoods and need to necessarily guide the sustainable livelihoods approach (SLA).

Box 4.1 The DFID Sustainable Livelihood Principles
Poverty-focussed development activity should be:

- **People-centred:** sustainable poverty elimination will be achieved only if external support focusses on what matters to people, understands the differences between groups of people and works with them in a way that is congruent with their current livelihood strategies, social environment, and ability to adapt.
- **Responsive and participatory:** poor people themselves must be key actors in identifying and addressing livelihood priorities. Outsiders need processes that enable them to listen and respond to the poor.
- **Multi-level:** poverty elimination is an enormous challenge that will only be overcome by working at multiple levels, ensuring that micro-level activity informs the development of policy and an effective enabling environment and that macro-level structures and processes support people to build upon their own strengths.
- **Conducted in partnership:** with both the public and the private sector.
- **Sustainable:** there are four key dimensions to sustainability—economic, institutional, social, and environmental sustainability. All are important—a balance must be found between them.
- **Dynamic:** external support must recognize the dynamic nature of livelihood strategies, respond flexibly to changes in people's situation, and develop longer term commitments.

Source: Ashley and Carney (1999)

4.3 Sustainable Livelihoods Framework

In 1998, IDS constructed an analytical framework to advance the understanding of sustainable rural livelihoods particularly in the context of the poor (Scoones, 1998; DFID, 1999). This framework is comprised of five elements, namely the 'vulnerability context', 'asset pentagon', 'policies, institutions and processes', 'livelihood strategy', and 'livelihood outcomes' that are interlinked as depicted in Fig. 4.1. The 'vulnerability context' comprises the external environment that impacts people comprising of shocks that are sudden, seasonality that is recurring, and trends that are gradual. This part of the framework lies furthest outside people's control and cannot be changed (DFID, 1999). The SLA identifies five core asset categories or the 'asset pentagon' on which livelihoods are built, namely human, social, natural, financial, and physical (Box 4.2). The access to assets increases from zero at the centre to a maximum at the perimeter of the pentagon. Asset endowments vary across households and are constantly changing and are depicted by differently sized pentagons and with arrows that depict the time dimension (Fig. 4.2). This framework can be used in planning and development for both assessing existing livelihoods as well as planning new ones.

Box 4.2 The Capital Asset Categories
Human capital represents skills, knowledge, education, capacity to work, and good health.
Social capital comprised of networks, groups, and relationships or the social resources upon which people depend to strengthen their livelihoods.
Natural capitals are the natural resources such as land, crops, livestock, forests, fisheries, water, ecosystem services, etc.
Physical capital includes infrastructure such as roads, transport, vehicles, shelter, water supply, energy, communication, irrigation, and the like.
Financial capitals are the financial resources that include income, savings, credit, debt, remittances, pensions, wages, etc.

The 'policies, institutions and processes' component comprises the policies of international organizations, national governments, and NGOs at micro, meso, and macro levels and the decision-making process, social norms, and customs and thereby determine the access to various capitals, terms of exchange between them, and the returns. A household decides its 'livelihood strategy' by combining the 'assets' it can access, taking into account its 'specific vulnerability context' and supported or obstructed by 'policies, institutions, and processes'. The execution of this 'livelihood strategy' results in 'livelihood outcomes' which may be positive or negative and result in either accumulation or atten-

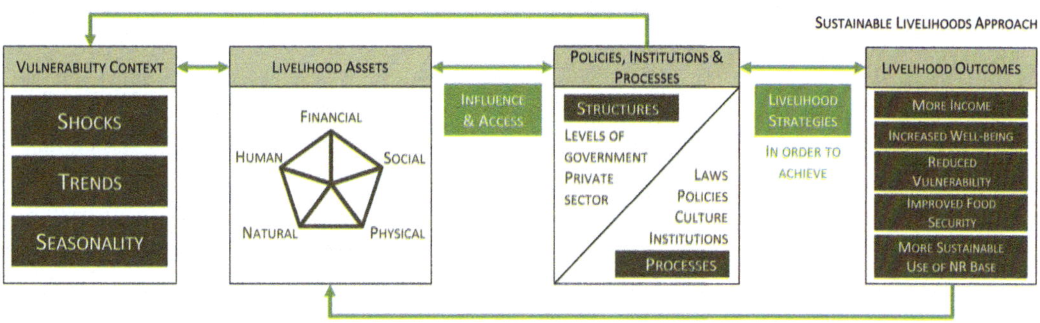

Fig. 4.1 The DFID sustainable livelihood framework (Source: Adapted from DFID, 1999, Donohue & Biggs, 2015)

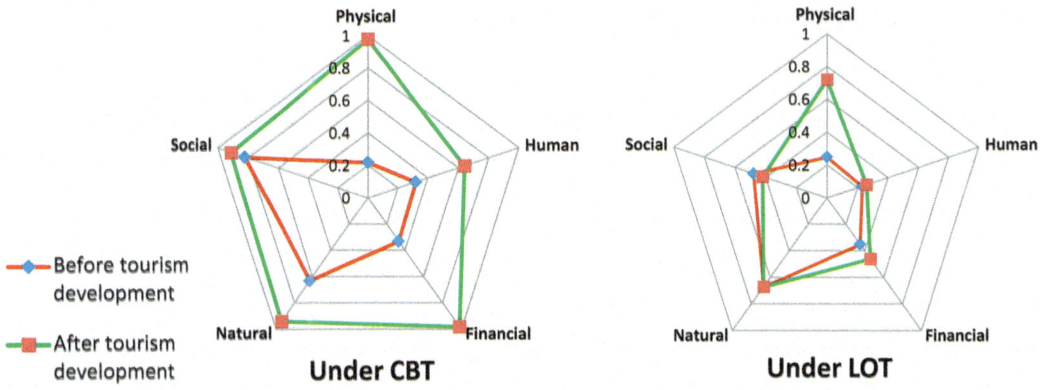

Fig. 4.2 The asset pentagon of the SLA can be quantified using indicators and indexing methods. A comparison of the two different governance models under community-based tourism (CBT) and lease-to-operate (LOT) in eastern China was done by quantifying the asset pentagon to obtain comparable indicators and facilitate meaningful interpretation (Source: Qian et al., 2017)

uation of its asset base. The livelihood outcomes are measured in terms of indicators such as income, food security, assets, wellbeing, vulnerability, and the like.

Assets and asset-building are at the core of any poverty reduction strategy. These assets function as a stock of resources that can be stored, exchanged, or allocated to activities to generate livelihood outputs and outcomes. The livelihood capitals can be accumulated as reserves and buffers to reduce vulnerability in times of stress or shocks (Scoones, 1998). According to Bebbington (1999), these assets are not merely the means for livelihood but also give meaning to a person's life and the capability to be and to act. Kim and Sumberg (2015) introduce the concept of assetness or the characteristics based on which assets can be categorized, such as their form, type, accessibility, fungibility, and use as detailed in Box 4.3.

Box 4.3 How Assets Are Categorized?

- **By form:** tangible (e.g. real property) and intangible (e.g. a claim);
- **By type:** physical (e.g. tools), natural (e.g. water, biodiversity), financial (e.g. savings), human (e.g. health, education, skills), social (e.g. relationships, networks);
- **By accessibility:** current (e.g. cash), deferred (e.g. insurance), and fixed (e.g. land);
- **By fungibility:** liquid (e.g. cash) or illiquid (e.g. an insurance policy);
- **By nexus of access or use:** individual, household, family, or community;
- **By form:** tangible (e.g. real property) and intangible (e.g. a claim);
- **By productivity or reproductivity:** performing (e.g. fruit-yielding trees) and non-performing (e.g. livestock in gestation).

Source: Kim and Sumberg (2015).

This publication was originally published by the Institute of Development Studies (IDS). You can access it at: Assets, 'Asset-ness' and Graduation | IDS Bulletin

The SLA saw widespread interest amongst donor agencies, researchers, and international development experts. There was an explosion of studies on livelihood assessment, monitoring, and evaluation in diverse geographies and settings such as pastoralism, agriculture, tourism, fishing, etc. SLA and PRA tools were integrated to make the whole process of livelihood assessment participatory and bottom-up. This SLA marked the shift from need-based, resource centred, agrarian solutions to an understanding of the complexities and diversity of livelihoods (Brocklesby & Fisher, 2003). The DFID SLA has been widely used by various organizations and adapted in various contexts, projects, and situations (Haidar, 2009). It provides a useful framework for analysing complexity, ensures that focus is not just on economic and direct impacts but also on the human and social aspects, and has an intuitive appeal for cross-sectoral and people-centred work (Ashley, 2000).

4.4 Critical Analysis

The main objective of SLA is to enhance the sustainability of livelihoods by building on the asset base of the poor households to enable them to take advantage of opportunities, reduce their vulnerability, and enhance their wellbeing (Foresti et al., 2007). The SLA is credited with a shift in development thinking from sectoral, top-down projects to bottom-up, household focussed investments (Mensah, 2012). It has been widely used in international development, however, over the last few decades its use has been receding from mainstream development practice. It is criticized as being too micro and too household focussed thereby limiting its ability to inform macro policy analysis or impact assessment (Table 4.1). Some feel that the framework has serious shortcomings: among these, the absence of political capital, gender, enterprise, markets, and other power issues and rights and also, the failure of the asset pentagon to distinguish between personal and common assets (Hussein, 2002). Dorward et al. (2003) argue about the need to add markets, institutions, and technology, explicitly in the framework.

Table 4.1 Strengths and weaknesses of the Sustainable Livelihoods Approach

Strength	Weakness
Emphasizes the importance of macro and micro linkages between assets, vulnerability context (which are local), and PIP (which is multi-level) in provisioning sustainable livelihoods	Underplays elements of the vulnerability context, such as macro-economic trends, national policies, international trade, conflict, etc.
Draws attention to the multiplicity of assets and moves beyond the focus on only monetary aspects. Seeks to understand changing configuration of the asset pentagon in provisioning different livelihood strategies in a dynamic and historical context	Assumes that capital assets can be expanded in a generalized and incremental fashion. Does not take into account the scenario where strengthening the livelihoods of one group can adversely impact those of the other
Highlights the need to move beyond the silo mentality and to look at livelihoods as a whole and not through narrow sectoral lenses	Undermines the role of power, inequality, discrimination, structural constraints, gender, and the difference between personal and common pool resources
Useful for evaluation such as efficacy studies, process evaluation, and the like	Assessing complex livelihood elements at the household level using qualitative tools will need a large number of qualified, frontline staff and hence is often expert-driven
Aggregated studies of homogenous groups such as wellbeing classes (well off, middle class, poor) or livelihood categories (weavers, shrimp farmers, forest dwellers, etc.) can be handy for project management or informing policy	Livelihood studies typically taken up at the household level are highly contextualized to aid in generalization, macro policy formulation, or to challenge existing theories. Project planners desire an aggregated view to gain insights on livelihood strategy and project components
Useful in theorizing and constructing the theory of change of a project or in livelihood assessment at the micro level	Difficult to integrate into a livelihood proposal as the constituent elements are abstract, complex, and dynamic

Source: Adapted from Hussein, 2002, de Haan 2012, Scoones, 2015, Serrat, 2017, Quandt, 2018

4.5 Innovations in Research and Practice

From a practitioner's perspective, where the requirement is a policy analysis and informing planning, monitoring, and evaluation of development projects, the SLA does not automatically lend itself to ease in use. The key questions concerning the applicability of the SLA in various scenarios include: can the approach be used to provide macro-level generalizations that are more amenable for policy-making by assessing livelihoods of groups instead of households? Can it be used to assess the livelihood status of geographical regions so that the findings can help inform priority regional policies? Is the SLA largely an abstract, qualitative framework or certain elements such as the asset pentagon can be quantified so that it can be used for comparative analysis? Other than livelihood assessment which can inform project planning, can the SLA be integrated with other components of project management such as monitoring and evaluation? We attempt to answer these questions by providing the gist of select research studies that explore innovative methods to address these perceived weaknesses of the SLA and strengthen its applicability in practice.

Cramb et al. (2004) used the SLA in combination with PRA tools to assess rural livelihoods in the highlands of Vietnam. Participatory rural appraisal (PRA) is described by Chambers as '*a growing family of approaches and methods to enable local people to share, enhance and analyse their knowledge of life and conditions, and to plan, act, monitor and evaluate*' (Chambers, 1997, p. 102). The study found that while the SLA provided a suitable framework for monitoring and evaluation, PRA provided a suite of methods for involving farmers in the process. The participatory wealth ranking (PWR) PRA tool was used to segregate the diverse households into three broadly homogenous well-being categories, namely 'better off', 'average', and 'poor', which were then sampled using household interviews. The household-level results were then aggregated at the PWR category level and analysed using the SLA lens thereby providing insights on resource ownership and the resultant livelihood dynamics at the village level. This study is a good example of how a combination of SLA and PRA can be used to discern broad livelihood patterns to aid in policymaking.

Ellis and Freeman (2004) compare and contrast the rural livelihoods in four east African countries by taking up micro-level livelihood studies to inform macro-level poverty strategy. To capture the full range of livelihood circumstances, in each selected village a PRA wealth ranking exercise was carried out to identity three wealth categories. Within each of these strata, few households were sampled using household interviews. The study found that poor households in all countries have low land and livestock holding, dependence on food crop farming, and weak market support. The better-off households improved their wellbeing by diversifying their livestock ownership, engaging in non-farm self-employment, and income diversification. Based on this they propose the macro-level livelihood strategy to promote public services to facilitate non-farm enterprises, agriculture extension services for farmers, and ascertaining the necessity of tax revenue collection. This study provides clarity on how an in-depth, micro-level livelihood assessment can be used to develop a macro-level livelihood strategy.

Masud et al. (2016) used the SLA for a standard-of-living assessment of communities living within marine protected areas in Malaysia using mixed methods comprising of focus group discussions and household surveys. They found that while social and physical assets of the local communities improved with tourism development, but their human, financial, and natural capital continues to be weak. This study provides a useful framework on how the SLA can be integrated into impact evaluation.

While undertaking a livelihood dynamics study in Kenya, Kristjanson et al. (2010) based on the predominant livelihood characteristics, segregated five broad 'livelihood zones', namely high potential, marginal, agro-pastoral, pastoral, and urban districts. This study provides a useful basis for integrating SLA in country-level assessments by stratifying the geographical landscapes into livelihood strategy strata which can then be taken up under detailed micro-level livelihood assessment by embedding SLA and PRA tools. The findings of this assessment were then aggregated to form large-scale generalizations to form regionally differentiated poverty policies.

Qian et al. (2017) evaluated the before-and-after impact of tourism development on rural livelihoods across two different governance models under community-based tourism (CBT) in Jade valley and lease-to-operate (LOT) in Phoenix valley in eastern China. They quantified the asset pentagon of the SLA using indicators and adopted scaling and indexing methods to obtain comparable indicators and facilitate meaningful interpretation (Fig. 4.2).

Belcher et al. (2013) developed an indicator-based livelihood monitoring tool using SLA by developing indicators to represent the five capitals. To calibrate and test the tool, they collected data for the indicators in six villages. The study found that while it was easier to assess the physical and financial assets, the social and natural capital were more difficult to quantify.

Donohue and Biggs (2015) highlight the need for effective monitoring of sustainable development at the sub-national level in Nepal to better understand the spatial variation in factors that contribute to sustainable livelihoods. Based on the SLA, the asset pentagon is quantified using 23 socio-environmental indicators to construct the multidimensional livelihoods index (MLI). The MLI was applied to subregions within Nepal to identify spatial variations to determine key policy priority concerns (Fig. 4.3).

With rising levels of crisis and shocks, the concept of livelihood resilience is gaining prominence in international development. Quandt (2018) developed a framework to measure household livelihood resilience by building on the asset-based approach of SLA. A total of 25 quantitative indicators were developed around the five capitals and aggregated to construct the composite asset index. The visualization was done with the help of radar diagrams or spider graphs to enable comparative analysis between different groups. This study helped to highlight the importance of assets in building livelihood resilience, especially the inclusion of non-monetary measures such as social and physical capital, thus recognizing the role that power relations can have in access to assets (Quandt, 2018).

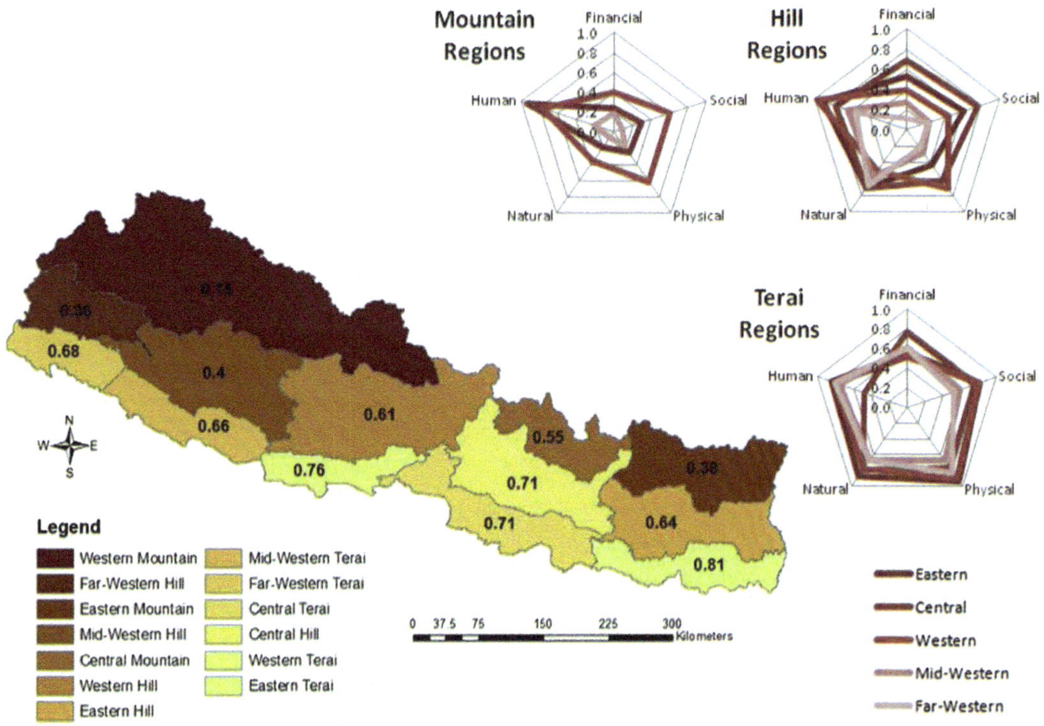

Fig. 4.3 The asset pentagon was quantified across the 13 sub-regions of Nepal using 23 socio-environmental indicators to construct a comprehensive livelihood index. This study identified spatial variations in the livelihood status and helped inform policy priority areas (Source: Donohue & Biggs, 2015)

4.6 The Way Forward

The SLA is credited with influencing the development paradigm from a top-down, sectoral focus to a bottom-up, household-centric motivation. Being abstract, complex, and qualitative its application was largely expert-led. While there was an explosion of studies that used this approach, they were highly contextualized and hence had limited influence on macro policy formulation or in challenging existing theories. However, some studies as discussed above creatively combined the SLA with PRA tools and quantitative techniques to enable comparative analysis, regional assessment, policy analysis, and impact evaluation. This helped enhance the applicability of the SLA in livelihood assessment, project planning, monitoring, and evaluation. Scoones (2015) argues that there is a need to broaden the scope of this approach by including political capital in the pentagon, engaging with politics and power, and the requirement to move across scales from micro to macro. Also, to make the SLA relevant again in international development there is a need to embed it in contemporary themes such as resilience studies, vulnerability assessments, urban studies, rights-based approach, etc. The SLA can emerge as an important tool in the present context to examine the impact of the COVID-19 pandemic on the livelihoods of vulnerable rural households around the world. The SLA can be very useful when integrated with other relevant approaches that are contextually suitable. The results from such studies can go a long way in influencing policy to safeguard the progress made on important SDGs like ending poverty and hunger.

Exercises

1. An indigenous forest dependent community in central India practising subsistence-level farming and foraging was translocated from deep inside a tiger reserve to a location far away from the forest having proximity to a nearby town. They were provided better housing, electricity, water supply, sanitation, access to schools, health facilities, and irrigated farmland. Illustrate how the asset polygon of this community will transform in this new location compared to the forest existence earlier?
2. Kabra (2009) assessed the livelihood outcomes from conservation induced displacement in two protected areas of India. You may like to watch the YouTube video provided below to get an immersion into the lives of the Sahariya Tribes—a Particularly Vulnerable Tribal Group (PVTG).

   https://www.researchgate.net/publication/45260695_Conservation-induced_Displacement_A_Comparative_Study_of_Two_Indian_Protected_Areas

   https://www.youtube.com/watch?v=njHrUo-OXko

   (a) Use the DFID sustainable livelihoods approach to illustrate how the asset polygon of the two communities changed before and after relocation?
   (b) Build on the DFID sustainable livelihoods approach to discuss the key assets that the local community in Bhadra possessed that enabled them to get a better relocation deal compared to the Sahariya's in Kuno?

3. Bebbington (1999) critically analyses the rural development approaches in the Andes and highlights the need to assess the different assets that people access and how they mix and convert these assets to build their livelihoods, the pathways they adopt to expand their asset base by engaging with other stakeholders, and how they are able to enhance their capabilities with particular focus on the social capital. Based on this paper, discuss the drawbacks in the conceptualization of the rural development programmes in the Andes and how they can be strengthened by using the sustainable livelihoods approach.

   http://citeseerx.ist.psu.edu/viewdoc/download?doi=10.1.1.589.495&rep=rep1&type=pdf

4. Study the journal paper by Ahmed et al. (2008) and also get an immersion into the context of 'Gher farming in Bangladesh' from the video before proceeding to the five questions listed below. Video: https://www.youtube.com/watch?v=mBvnbYgheDg
    - How did the authors develop the 'vulnerability context'?
    - How was the 'asset pentagon' constructed? Can it be represented as a radar diagram or a spider graph?
    - How were the 'transforming structures and processes' documented?
    - What methods were used to assess the 'livelihood strategy and outcomes'?
    - Do you think the DFID SLA provides a holistic insight into the livelihoods of the Gher farmers? Give reasons for your answer.
5. Carrying out livelihood assessment on large scale may need segregating the study area into livelihood zones as the household assets, activities, and outcomes differ. Kristjanson et al. (2010) carried out a country-wide poverty dynamics study in Kenya using the stages-of-progress methodology. Analyse how the livelihood strategies differed across these zones and the need for regionally differentiated poverty policies.
   https://sites.duke.edu/krishna/files/2013/06/Understanding-Poverty-Dynamics-in-Kenya.pdf
6. The Sustainable Livelihoods Approach (SLA) was conceptualized by Ian Scoones about two decades back. What improvements would you like to suggest to this approach? What are your views on the relevance of SLA in contributing to the SDGs global developmental agenda 2030?
   https://www.youtube.com/watch?v=nXUo0pEgKJw
7. Study the journal paper by Ellis and Freeman (2004) where they compare and contrast the rural livelihoods in four east African countries. Discuss the asset-based approach they use for participatory wealth ranking, the causes of poverty, and the implications of the study to inform evidence-based policymaking.
   http://oar.icrisat.org/1285/1/JouOfDevStu40%284%291-30__2004.pdf
8. Study the journal paper by Bryceson (2002) and discuss the livelihood change sweeping across the rural areas of sub-Saharan Africa and the factor driving this change.
   https://www.researchgate.net/profile/DeborahBryceson/publication/222046640_The_Scramble_in_Africa_Reorienting_Rural_Livelihoods/links/5a04b757458515eddb806a30/The-Scramble-in-Africa-Reorienting-Rural-Livelihoods.pdf
9. Assets and asset-building are at the core of any poverty reduction strategy. Study the journal paper by Kim and Sumberg (2015) where they introduce the concept of assetness—the features and characteristics of different assets. Discuss how this concept can help in better designing poverty reduction programmes especially in the context of livestock.
   https://opendocs.ids.ac.uk/opendocs/bitstream/handle/20.500.12413/7276/IDSB_46_2_10.1111-1759-5436.12135.pdf?sequence=1
10. Livelihood assessment is often a prerequisite before planning any livelihood intervention. Cramb et al. (2004) took up a participatory assessment of rural livelihoods in the central highlands of Vietnam. Discuss how the study used a combination of PRA techniques and the DFID sustainable livelihoods approach to take up a rapid and cost effective livelihood assessment.
11. Using a combination of participatory methods and sample surveys helps to achieve a more in-depth understanding of rural livelihood strategies. Malleson et al. (2008) investigated the rural livelihoods of people living in the forest zone of West and Central Africa especially in terms of the contribution of non-timber forest products to their livelihoods. Analyse the sampling design and methods they used, what practical difficulties they faced, and how the learnings can help in improving the design of future research.
    https://digitalcommons.unl.edu/cgi/viewcontent.cgi?article=1026&context=icwdmeea
12. How does one embed the sustainable livelihoods approach (SLA) in an impact evaluation? Reddy et al. (2004) use this framework to assess the impact of a watershed development programme on rural livelihoods. Discuss the methodology they used and how the learnings from this evaluation can help to better design future watershed programmes.

https://rmportal.net/framelib/watershed-devt-in-india.pdf

13. How does one use the DFID sustainable livelihoods approach for a livelihood assessment? Masud et al. (2016) used this framework to undertake a comprehensive livelihood assessment of the Marine National Park area in Malaysia. Analyse the methodology used and how a combination of survey and participatory approaches was used including the sampling design.
    https://link.springer.com/article/10.1007%2Fs11205-015-0872-2
14. How does one quantify the asset pentagon of the DFID sustainable livelihoods framework? Qian et al. (2017) evaluate the impact of tourism development on rural livelihoods across two different governance models in eastern China. Critically analyse the methodology used for asset quantification and the before-and-after impact assessment carried out by them.
    https://www.sciencedirect.com/science/article/pii/S0261517717300067
15. Livelihood strategies build on the asset pentagon. Study the research carried out by Hua et al. (2017) in the eastern Tibetan plateau and discuss the methodology they used to develop the asset pentagon and also ascertain its relationship with the livelihood strategies adopted.
    https://www.sciencedirect.com/science/article/pii/S1470160X17301176
16. Donohue and Biggs (2015) advocate the use of a multidimensional spatial approach to monitor social and environmental change at the sub-national level in rural Nepal. They selected 23 indicators to construct a multidimensional livelihood index (MLI) and find considerable spatial variability in the factors affecting people's livelihoods and thereby highlight the potential locations and livelihood strategies needed that can help in the targeting of resources. Comment on the usefulness of the MLI tool to monitor livelihood wellbeing and to assess geographical variability across regions as a decision-support framework.
    https://doi.org/10.1016/j.apgeog.2015.05.006
17. The asset pentagon of the SLA was quantified using 25 indicators and aggregated to develop a composite asset index to measure household livelihood resilience by Quandt (2018). Comment on the methodology adopted and the usefulness of this approach.
    https://www.researchgate.net/profile/Amy-Quandt/project/Building-Livelihood-Resilience-to-the-Impacts-of-Climate-Change-What-Role-does-Agroforestry-Play/attachment/5ad604e34cde260d15d94fb3/AS:616411638534145@1523975395381/download/World+Development.pdf
18. Ludi and Slater (2008) analyse the DFID sustainable livelihoods approach to assess its strengths and weaknesses. Based on this analysis, discuss what improvements you would like to propose to the original framework.

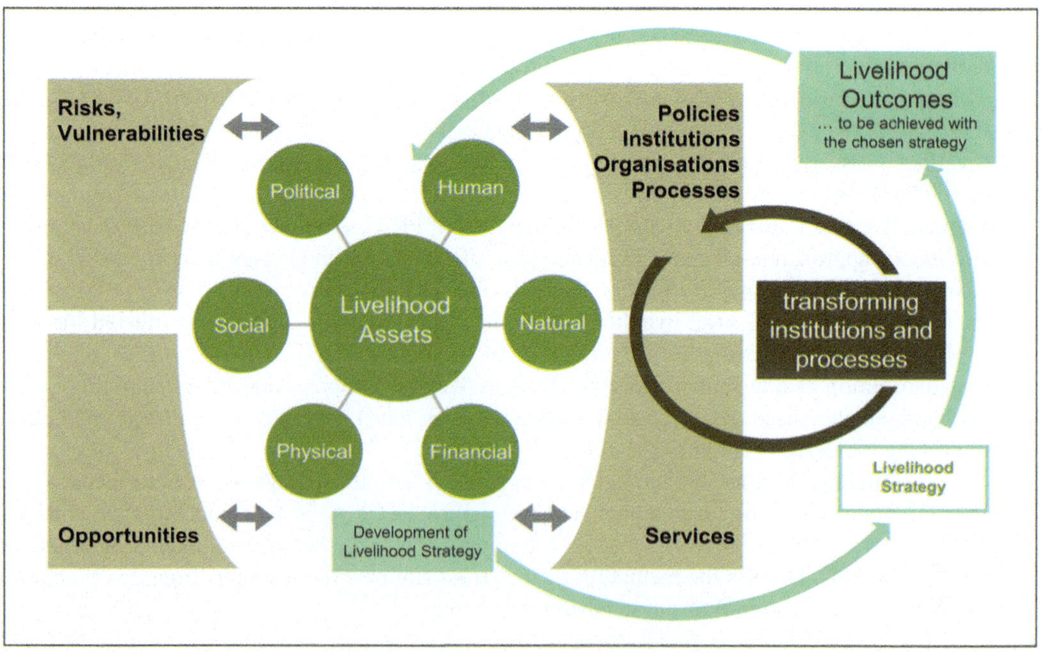

https://www.shareweb.ch/site/PovertyWellbeing/resources/Documents/Briefing%20Note%2011%20%20Using%20the%20Sustainable%20Livelihoods%20Framework%20to%20Understand%20and%20Tackle%20Poverty.pdf

19. The Sustainable Livelihood Approach was conceptualized by IDS researcher Ian Scoones about two decades back. What improvements would you like to suggest to this approach? Listen to his podcast as he discusses his book 'sustainable livelihoods and rural development' to supplement your answer. https://youtu.be/-d3KoxQIlcs
20. What are your views on the relevance of Sustainable Livelihoods Approach in contributing to the SDGs global developmental agenda 2030? Listen to his podcast by IDS researcher Ian Scoones as he discusses his book 'sustainable livelihoods and rural development' to supplement your answer. https://youtu.be/-d3KoxQIlcs

References

Ahmed, N., Allison, E. H., & Muir, J. F. (2008). Using the sustainable livelihoods framework to identify constraints and opportunities to the development of freshwater prawn farming in Southwest Bangladesh. *Journal of the World Aquaculture Society, 39*(5), 598–611.

Ashley, C. (2000). *Applying livelihood approaches to natural resource management initiatives: Experiences in Namibia and Kenya*. Overseas Development Institute.

Ashley, C., & Carney, D. (1999). Sustainable livelihoods: Lessons from early experience (7) , 1. : Department for International Development.

Bebbington, A. (1999). Capitals and capabilities: A framework for analyzing peasant viability, rural livelihoods and poverty. *World Development, 27*(12), 2021–2044.

Belcher, B., Bastide, F., Castella, J. C., & Boissière, M. (2013). Development of a village-level livelihood monitoring tool: A case-study in Viengkham District, Lao PDR. *International Forestry Review, 15*(1), 48–59.

Brocklesby, M. A., & Fisher, E. (2003). Community development in sustainable livelihoods approaches–an introduction. *Community Development Journal, 38*(3), 185–198.

Bryceson, D. F. (2002). The scramble in Africa: reorienting rural livelihoods. *World development, 30*(5), 725–739.

Carney, D. (1998). Implementing the sustainable livelihoods approach. In D. Carney (Ed.), *Sustainable rural livelihoods: What contribution can we make?* (pp. 3–23). Department for International Development (DFID).

Carney, D. (2003). *Sustainable livelihoods approaches: Progress and possibilities for change* (p. 64). Department for International Development.

Chambers, R. (1997). *Whose reality counts? Putting the last first*. Intermediate Technology Publications.

Chambers, R., & Conway, G. (1992). *Sustainable rural livelihoods: Practical concepts for the 21st century*. Institute of Development Studies (UK).

Cramb, R. A., Purcell, T. C. S. H., & Ho, T. C. S. (2004). Participatory assessment of rural livelihoods in the central highlands of Vietnam. *Agricultural Systems, 81*(3), 255–272.

De Haan, L. J. (2012). The livelihood approach: A critical exploration. *Erdkunde*, 345–357.

DFID. (1999). *Sustainable livelihoods guidance sheets*. UK DFID, Department for International Development.

Donohue, C., & Biggs, E. (2015). Monitoring socio-environmental change for sustainable development: Developing a multidimensional livelihoods index (MLI). *Applied Geography, 62*, 391–403.

Dorward, A., Poole, N., Morrison, J., Kydd, J., & Urey, I. (2003). Markets, institutions and technology: Missing links in livelihoods analysis. *Development Policy Review, 21*(3), 319–332.

Ellis, F., & Freeman, H. A. (2004). Rural livelihoods and poverty reduction strategies in four African countries. *Journal of Development Studies, 40*(4), 1–30.

Foresti, M., Ludi E., & Griffiths, R. (2007). Human rights and livelihood approaches for poverty reduction, ODI Paper prepared for the SDC Poverty-Wellbeing Platform. Retrieved on 5th June 2021 at 10:45 am from https://cdn.odi.org/media/documents/HR_livelihood_approaches.pdf

Haidar, M. (2009). Sustainable livelihood approaches. In the framework, lessons learnt from practice and policy recommendations. Expert group meeting on adopting the sustainable livelihoods approach for promoting rural development in the ESCWA region Beirut. Retrieved on 14th June, 2021 at 10:00 from http://css.escwa.org.lb/sdpd/1125/undp.pdf

Hua, X., Yan, J., & Zhang, Y. (2017). Evaluating the role of livelihood assets in suitable livelihood strategies: Protocol for anti-poverty policy in the Eastern Tibetan Plateau, China. *Ecological Indicators, 78*, 62–74.

Hussein, K. (2002). *Livelihoods approaches compared*. Department for International Development.

Kabra, A. (2009). Conservation-induced displacement: a comparative study of two Indian protected areas. *Conservation and society, 7*(4), 249–267.

Kim, S. K., & Sumberg, J. (2015). Assets, 'Asset-ness' and Graduation. *IDS Bulletin, 46*(2), Brighton: Institute of Development Studies, 124–133. Assets, 'Asset-ness' and Graduation | IDS Bulletin.

Kristjanson, P., Mango, N., Krishna, A., Radeny, M., & Johnson, N. (2010). Understanding poverty dynamics in Kenya. *Journal of International Development, 22*(7), 978–996.

Ludi, E., & Slater, R. (2008). Using the sustainable livelihoods framework to understand and tackle poverty: Briefing note. Swiss Agency for Development and Cooperation. Retrieved on 11th June, 2021 at 21:00 from https://www.shareweb.ch/site/PovertyWellbeing/resources/Documents/Briefing%20Note%2011%20%20Using%20the%20Sustainable%20Livelihoods%20Framework%20to%20Understand%20and%20Tackle%20Poverty.pdf

Malleson, R., Asaha, S., Sunderland, T., Burnham, P., Egot, M., Obeng-Okrah, K., ... & Miles, W. (2008). A methodology for assessing rural livelihood strategies in West/Central Africa: lessons from the field. *Ecological and Environmental Anthropology (University of Georgia)*, 25.

Masud, M. M., Kari, F., Yahaya, S. R. B., & Al-Amin, A. Q. (2016). Livelihood assets and vulnerability context of marine park community development in Malaysia. *Social Indicators Research, 125*(3), 771–792.

Mensah, E. (2012). The sustainable livelihood framework: A reconstruction. *The Development Review – Beyond Research, 1*(1), 7–25.

Qian, C., Sasaki, N., Jourdain, D., Kim, S. M., & Shivakoti, P. G. (2017). Local livelihood under different governances of tourism development in China–A case study of Huangshan mountain area. *Tourism Management, 61*, 221–233.

Quandt, A. (2018). Measuring livelihood resilience: The household livelihood resilience approach (HLRA). *World Development, 107*, 253–263.

Reddy, V.R., Reddy, M.G., Galab, S., Soussan, J., & Springate-Baginski, O. (2004). Participatory watershed development in India: Can it sustain rural livelihoods? *Development and change, 35*(2), 297–326.

Scoones, I. (1998). *Sustainable rural livelihoods: A framework for analysis. Working paper 72.* Institute for Development Studies.

Scoones, I. (2015). *Sustainable livelihoods and rural development*. Practical Action Publishing.

Sen, A. (1982). *Choice, welfare and measurement.* Blackwell; and Cambridge Mass: MIT Press.

Sen, A. (1985). Well-being, agency and freedom: The Dewey Lectures 1984. *Journal of Philosophy, 82.*

Sen, A. (1993). Capability and Well-being. In A. K. Sen & M. Nussbaum (Eds.), *The quality of life* (pp. 30–53). Oxford University Press.

Serrat, O. (2017). The sustainable livelihoods approach. In *Knowledge solutions* (pp. 21–26). Springer.

Small, L. A. (2007). The sustainable rural livelihoods approach: A critical review. *Canadian Journal of Development Studies/Revue canadienne d'études du développement, 28*(1), 27–38.

Solesbury, W. (2003). Sustainable livelihoods: a case study of the evolution of DIFD policy. Overseas Development Institute, Working Paper 217, 1–36.

Chapter 5
Commons and Livelihoods

Abstract Commons such as forests, pastures, fisheries, irrigation, and others play a vital role in sustaining the livelihoods of the poor. These are open access resources that can be easily over-consumed resulting in their depletion. Often commons property regimes exist where institutions frame formal rules along with informal codes to regulate access to the resource. Empirical studies show that higher levels of local enforcement are associated with sustainable outcomes measured as better health of commons and livelihood benefits. In many long enduring commons, the state adopts a proactive role to provide strategic support, without appropriating ownership or decision-making rights. While there is a lot of optimism associated with community-based management, in reality, it is complex and raises questions of social difference, conflicting concerns, power structures, etc. As a result, practical implementation of community management is challenging and has not always been successful. A polycentric management of the commons where the state provides strategic investment without participating in ownership of the commons or its management has the best chance of achieving sustainable outcomes. Considering the significant contribution of the commons in sustaining the livelihoods of the poor, we opine that they deserve greater attention in the policies directed towards ending poverty and hunger.

Keywords Ostrom · Excludable · Rivalrous · Open access · Resource system · Resource unit · Appropriation · Design principles · Governance · Rule making · Collective action · Collective-choice · Polycentric

5.1 Nature of Common Environment Goods

In economics, four types of goods are defined, namely private goods, common goods, club goods, and public goods, based on their excludability and rivalrousness in consumption (Ostrom, 1990; Araral, 2014) (Fig. 5.1). A rivalrous good is one whose consumption by one consumer prevents simultaneous consumption by other consumers such as cars, mobiles, timber, fisheries, etc. A good is excludable when it is possible to prevent consumers, who have not paid for it from having access to it such as cinemas, museums, etc. Private goods are excludable and rivalrous such as cars, mobile phones, etc. Club goods are excludable but non-rivalrous and require a 'membership' payment to enjoy the benefits of the goods such as cinemas, private clubs, private parks, etc. Common goods are non-excludable and rivalrous and hence are easily over-consumed. Public goods are non-excludable and non-rivalrous, people cannot be excluded from using them, and use by an individual does not reduce the availability to others such as public parks, the air we breathe, peace in our neighbourhood, etc.

Common Pool Resources (CPRs) are common goods—man-made or natural, having open access to resources. They are for the most part rivalrous, as one person's use of the commons subtracts from its use by others, and there are non-excludable as the nature of the resource is such that it is difficult to exclude access (Ostrom, 1990). These two characteristics of being non-excludable and rivalrous

S. Tambe, *Teaching and Learning Rural Livelihoods*, Sustainable Development Goals Series,
https://doi.org/10.1007/978-3-030-90491-3_5

		SUBTRACTABILITY	
		Low	*High*
EXCLUSION	*Difficult*	**Public goods** Air, useful knowledge, peace sunsets, national defence	**Common goods** Forests, fisheries, pastures, public libraries, Irrigation systems
	Easy	**Club goods** Journal subscriptions, cinemas day-care centres, private parks, museums, satellite TV	**Private goods** Food, clothing, personal computers, mobiles

Fig. 5.1 There are four types of goods in economics, based on whether these goods are excludable and rivalrous in consumption (Source: Ostrom & Ostrom, 1977)

make CPRs susceptible to overharvesting and destruction (Araral, 2014). Some common examples of commons include forests, fisheries, irrigation, pastures, etc. Ostrom (1990) describes the resource dynamics of a CPR succinctly as follows. A CPR can be natural or man-made, and in this chapter, we focus on natural CPRs which are large enough to make it costly to exclude them from potential users. A CPR comprised of 'resource system' which is a stock variable that under favourable conditions produces a finite flow of 'resource units'. For example, fishing grounds, groundwater resources, pastures are resource systems, while the tonnes of fish harvested, cubic metres of groundwater extracted, or tonnes of fodder consumed, etc. are the resource units. Resource units are what users withdraw from the resource systems, and this action is termed as appropriation and those who appropriate are called appropriators such as herders, fishers, irrigators, etc. While the actions of the appropriators affect the resource system and the yield of resource units, organizing them for collective action is a complex task. What makes collective action complicated is that common pool resource are rivalrous or subtractable in nature, that is over-extraction of the resource units may impair the capability of the resource system to continue producing them in the future.

5.2 Contribution of the Commons to Rural Livelihoods

In developing countries, environmental incomes contribute about a quarter of the total household income, three-fourths of which is contributed by natural forests (Angelsen et al., 2014). The authors go on to highlight that the share of environmental incomes is higher for low-income households sourced mainly from subsistence products such as fuelwood and wild foods. In the dry regions of India, the CPRs contribute significantly to household incomes, at times even more than the poverty reduction programmes themselves (Jodha, 1986). Also, the contribution of the CPRs is most notable in the vulnerable villages located furthest from the road head and markets, where the households depend on the forests for the majority of their income and livelihood requirements (Ghate et al. 2009). Forests serve as a safety net for forest fringe communities by provisioning 'famine foods', health (medicinal plants), and cash income (from the sale of forest products) (Wunder, 2001). Also, poor households derive a larger share of their income from forests and wildlands than better-off households within the same community. The rights of the poor people to the forests tend to be open or informal and difficult to protect against competing interests such as logging firms, commercial farmers, ranchers, mining companies, and the like (Wunder, 2001).

5.3 The Tragedy of the Commons

Common pool resources can be overexploited because they are subtractable and non-excludable leading to a phenomenon known as 'tragedy of the commons'. Hardin (1968) in his seminal paper described how each user of the commons would act to maximize their benefits from the open access commons, while the costs of their use were shared between all users. As a result, the commons would be subject to overuse and this would eventually lead to degradation and collapse of the resource. For instance, a pasture can sustain a particular stocking density or the number of livestock grazing on it every year, without degrading the core stock resource. But if the pasture is overgrazed, it will start degrading and its productivity will get impaired. As CPRs are finite, rivalrous, and non-excludable, they are often subjected to pressures beyond their carrying capacity resulting in reduced productivity and eventual destruction unless extraction is regulated and monitored locally. Degradation of the commons fuels a vicious cycle of resource depletion. Increased livestock density results in a decrease of pasture and water available per animal, resulting in a decline in livestock health (SAPPLPP, 2009). This degradation of the CPRs manifests itself in the form of soil erosion, soil nutrient depletion, moisture stress, deforestation, biodiversity loss, fodder scarcity, etc., and thus negatively impacts farming (SAPPLPP, 2009). At times, weak management of state-owned forests results in open access with resultant overuse and degradation (Quinn et al., 2007). In India like in many other parts of the global south, the area and productivity of the CPRs have declined over the years and they deserve greater policy attention, as when the commons degrade commons-dependent poor become poorer (Jodha, 1986).

5.4 Common Property Regimes

CPRs are not always open access and common property regimes for collective governance may exist with local rule making dictating the access and usability of the resource for the users. When rights and duties are adequately enforced through common property regimes, CPRs are not always subject to open access and degradation (Cousins, 2000). Ostrom's pioneering work on the commons challenged the notion that rational human beings would be driven only to maximize their own short-term gains (Frischmann, 2013). She challenged the general notion that to avoid 'tragedy of the commons', the only option is to hand over the governance to either private firms or the state (Quinn et al., 2007). Instead, she suggested that self-governance can be a viable alternative to address the challenge of collective action and is based on the centrality of trust and reciprocity (Ostrom, 1998) (Box 5.1).

CPR theory establishes conditions under which institutions will work best and specifies 'design principles' which include the need for clear resource boundaries, relative socio-economic homogeneity, sanctions, rules, monitoring, and so on (Wade, 1988; Ostrom, 1990) (Box 5.2). The CPR theory presents the potentials and possibilities of collective action in natural resource management (Mehta et al., 2001). To reduce the cost of peer-to-peer monitoring, these institutions frame formal rules along with informal codes to minimize the cost of constant monitoring of user behaviour and are thus efficient (Box 5.3). Developing functional and legally recognized CPR regimes can be politically, socially, and institutionally challenging and the scholarship on design principles can assist in this task (Levine & Richmond, 2015). These conditions are however often not easy to recreate and are more apt for long standing CPRs. So then how to go about establishing a common property regime adhering to these 'design principles'? Morrow and Hull (1996) share the difficulties in creating successful CPR institutions by external actors and suggest that Ostrom's eight design principles are apt for long standing CPRs. They further add that external projects often fail to realize the need to assume a facilitator's role

Box 5.1 Elinor Ostrom—Commons Champion

Elinor Ostrom, née Elinor Claire Awan was born in Los Angeles, California, and grew up in a family of modest means. She was the first one in her immediate family to attend college. She studied political science at the University of California Los Angeles, where she also earned a Ph.D. in 1965. She later went on to work at Indiana University in Bloomington and was also associated with Arizona State University in Tempe and Virginia Tech. In her formative years, Ostrom was fond of swimming as a recreational activity and was an active member of her school debate team. Debating had a profound impact on her way of thinking as it involved observing problems from all possible angles and critiquing both sides.

In her graduation work, she researched the water sector in Southern California. It was during this time that she had her first exposure to studying common pool resources and understanding the problems associated with their management of such resources. It was 15 years post her graduation that she turned once again to studying the commons. At that time, the conventional wisdom was that natural resources that were collectively managed would eventually get degraded by overuse. Ostrom synthesized empirical data from around the world and also conducted field studies and showed that local communities in diverse settings such as fisheries, forests, pastures, etc. have been successfully managing the commons. They have evolved rules for collective action leading to sustainable use of natural resources.

Ostrom's work has led to a renewed global interest in the commons and their governance. In 2009, she was awarded the Nobel Memorial Prize in Economic Sciences for her 'analysis of economic governance, especially the commons'. She is the first woman to have won a Nobel Prize in Economics.

Source: https://www.nobelprize.org/prizes/economic-sciences/2009/ostrom/biographical/

that catalyses the self-organization of the CPR institutions, rather than imposing a readymade system from outside (Box 5.4). Measures to facilitate local self-government can be impeded by power asymmetries and distrust amongst users that may constrain the evolution of social capital and collective action (Theesfeld, 2004). In community-based forest management, there is a risk of elite capture of forest harvest benefits, which is substantially moderated when external organizations are involved (Persha & Andersson, 2014). Also, when we look at larger sized commons like seas, rivers, air, and groundwater, their status, and the problems they are plagued with, one wonders whether the CPR theory applies to only small and medium sized commons. Araral (2014) suggests that while Ostrom's findings hold for small-sized commons, but for large-scale commons at the national or global level Hardin is more justified.

Box 5.2 CPR Design Principles (Source Ostrom, 1990; Cox et al., 2010)

- **1A User boundaries:** Clear boundaries between legitimate users and nonusers must be clearly defined.
- **1B Resource boundaries:** Clear boundaries are present that define a resource system and separate it from the larger biophysical environment.
- **2A Congruence with local conditions:** Appropriation and provision rules are congruent with local social and environmental conditions.
- **2B Appropriation and provision:** The benefits obtained by users from a common pool resource (CPR), as determined by appropriation rules, are proportional to the amount of inputs required in the form of labour, material, or money, as determined by provision rules.

- **3 Collective-choice arrangements:** Most individuals affected by the operational rules can participate in modifying the operational rules.
- **4A Monitoring users:** Monitors who are accountable to the users monitor the appropriation and provision levels of the users.
- **4B Monitoring the resource:** Monitors who are accountable to the users monitor the condition of the resource.
- **5 Graduated sanctions:** Appropriators who violate operational rules are likely to be assessed (depending on the seriousness and the context of the offense) by other appropriators, by officials accountable to the appropriators, or by both.
- **6 Conflict-resolution mechanisms:** Appropriators and their officials have rapid access to low-cost local arenas to resolve conflicts among appropriators or between appropriators and officials.
- **7 Minimal recognition of rights to organize:** The rights of appropriators to devise their own institutions are not challenged by external governmental authorities.
- **8 Nested enterprises:** Appropriation, provision, monitoring, enforcement, conflict resolution, and governance activities are organized in multiple layers of nested enterprises.

Box 5.3 Governing the Commons—Collective Action at Pachgaon Village in Central India

Pachgaon is a small village in Chandrapur district in the state of Maharashtra in India with 72% tribal population. In 2012 after a long struggle it received Community Forest Resource Rights (CFR) over 1006 hectares of forests with sizeable bamboo resources under the Forest Rights Act, 2006. Less than half of the households in the village owned cultivable land and their main source of income was wage labour and migration. The unique feature of this village is collective decision making in the *gram sabha* (village assembly). Attendance in the gram sabha meetings is compulsory, and there is a fine of Rs 50 for each missed meeting. As a first step after receiving CFR, all households contributed five regulations for forest protection. Of the total 500 regulations compiled, the gram sabha discussed and finalized 115 rules. They established a patrolling system, reporting cases of wood cutting, forest fire, and restricting access to outsiders. The wage of Rs 266 per day is provided for patrolling, maintaining fire line, soil moisture conservation work, etc. Absence from patrolling for unexplained reasons carried a fine of up to Rs 200. They enforced a graded system of fines for stealing from the forest—Rs 500 for people on motorcycles; Rs 300 for people on bicycles, and Rs 150 for those on foot and carrying a head-load. They set aside 34 ha as *Devrai* or sacred grove where no extraction of resource was permitted.

Building on their earlier experience with the forest department, they followed the standard 3-year bamboo felling cycle with one-third of the area is harvested every year. Thus the bamboo clumps got a three-year rest period to regenerate between the two harvests. The gram sabha decides the wage rate, the upper cap on harvesting, supervisors for different activities, expected selling price, etc. It appointed 38 persons in charge of overall management, record keeping, and maintaining accounts. Men and women both harvest bamboo and are paid equal wages. The quantum of bamboo harvested per person per day is capped at 55 long bamboos and 5 bundles, with a piece rate of Rs 6.78 per bamboo and Rs 20 per bundle. This translates to a wage of Rs 473 per day during the harvesting season which lasts for 4–6 months in a year. The gram sabha retains 10% of wages, which is released in the monsoon months when there is no bamboo cutting.

(continued)

Box 5.3 (continued)

Bamboo is stored in the village depot, where it is graded and sold through open auction. At the end of the bamboo harvest, disclosure is made on the quantity harvested, amount deposited in the bank, and investment plan of the village development fund. Over the last few years the village earned about Rs 55 lakh annually, half of it was spent on wages with the remaining contributing to the village development fund. Consequently, migration from the village in search of labour has reduced significantly.

Sources: Tambe et al., 2019, TISS, 2018

Box 5.4 Duped by Traders—The Story of Mirkal Village in Central India

In 2016, 600 villagers from Mirkal village in Gadchiroli district of Maharashtra state of India and 26 adjacent villages got busy harvesting bamboo in three forest compartments. Like many other villages in the district, the *gram sabha* (village assembly) had been provided autonomy to harvest and sell bamboo and other non-timber forest products. A trader from an adjacent town had promised them lucrative rates for bamboo bundles. The gram sabha agreed to this deal worth about Rs 149 lakh and the agreement was signed and formalized. The trader made payments in instalments totalling Rs 61 lakh in cash till half of the bamboo was lifted. He then wanted to lift the remaining material but was restrained and asked to first make the full payment. Villagers then got into an argument and there was a difference in opinion. The dominant group reasoned that unless the total material was lifted and sold, how would the trader make the balance payments. Hence, the trader was eventually permitted to lift the total material, but after that, he never returned to make the payments. The villagers requested the police to trace him but to no avail. The balance wage payments amounting to Rs 88 lakh of 600 people are still pending. The villagers attempted to nab two of the trader's trucks, but he got wind of the plan and did not turn up. This experience has turned into a lifelong curse for the Mirkal *gram sabha* (village assembly) president. He is continuously hounded by people and cannot leave the village or travel freely. This is an unintended consequence of simple and gullible villagers having to fend with wily traders without external support.

Sources: Tambe et al., 2019.

5.5 Drivers of Successful CPR Management

Factors such as institutional structure, collective-choice arrangements, local enforcement, resource type, resource size, resources provisioned, group heterogeneity, shared norms, social capital, local leadership, poverty levels, subsistence dependence, commercial dependence, the influence of external markets, diffusion of technology, the role of state and others all seem to play a key role as the crucial drivers of successful CPR management (Agrawal, 2001; Pagdee et al., 2006). Agrawal (2001) highlighted the complexity of the causal linkage and developed four clusters of variables, namely the resource system, user group characteristics, institution arrangements, external environment, and their interlinkages. He goes on to highlight the need for careful research design, sample selection, construction of causal mechanisms, and shift towards comparative and statistical approaches rather than single-case analysis. Chhatre and Agrawal (2008) analysed data from nine developing countries to show that local rule making, collective action, and monitoring were the key drivers of improving forest health. Small to medium sized forests with low levels of dependence had a higher chance of regeneration compared to larger forests, with higher levels of dependence. Forest systems are more likely to have sustainable outcomes measured as higher tree species diversity and livelihood dependence with local rule making, supported by larger forest size, and commercial livelihood dependence (Persha et al., 2011).

Even in developed countries like Japan, there are long enduring CPR institutions that are managing the irrigation systems successfully (Sarker & Itoh, 2001). The state despite having invested a lot in physical capital such as irrigation and drainage engineering works plays a non-coercive, strategic role and significantly contributes to irrigators' self-governance of their CPRs. While the government either partially or completely subsidizes the construction cost of the engineering projects, the operation and maintenance are customarily paid for by the local irrigation community regimes. In this model, while the provisioning of the construction costs (hardware) is met by the government's strategic investment, the management (software) part involving provisioning and appropriation to ensure that an irrigator with a larger area has to pay higher water fees is done by the irrigators. About present-day management of Japan's community-based coastal fisheries commons, the state authorities provide the statutory framework, financial support, and marketing facilities for the fishery cooperative organizations, but the authorities neither control the fisher's daily operations or management decisions (Sarker et al., 2015). Thus, though the state is substantially involved, its strategic position enhances self-governance and local autonomy. Hence, design principles 7 and 8 (Box 5.2) refer to the role of the state where it adopts a pro-user, self-governance role that provides strategic support, but neither takes ownership of the commons nor participates in the management or institutional arrangements (Sarker et al., 2015).

5.6 Implications for Sustainable Rural Livelihoods

CPRs play an important role in sustaining the livelihoods of the poor. As the poor lack personal assets, their dependence on these natural resources for their sustenance is much more. The commons under community-based governance do not always get degraded. Hence, the 'tragedy of the commons' argument should not be used to advocate for either state control or privatization. On the other hand, while there is a lot of optimism and promise in decentralized, community-based governance of the commons, practical implementation of community management has not always been successful (Leach et al., 1997). Thus, it would be apt to reframe the CPR narrative as 'drama of the commons' as there is evidence of both successful and failed institutions (Araral, 2014). Also, there is a need to look beyond the simplistic notions of community as bounded, homogenous, egalitarian, pro-poor, and neglecting questions of social difference, conflicting concerns, power structures, etc. (Mehta et al., 2001). Approaches need to address conflict rather than assume harmony, embrace complexity, contextual diversity, social inequality, and ecological heterogeneity, rather than assume uniformity, and work from an understanding of institutional diversity and dynamics (Leach et al., 1997). The role of the state is critical in ensuring polycentric governance by provisioning an enabling external environment. It needs to strategically contribute to policy, infrastructure, finance, and markets, while not sharing ownership of the commons or participating in local level management and decision making.

Despite the contribution of the CPRs to the economy, they have been relatively invisible in developmental policy and planning, which has emphasized private property resources (Jodha, 1986). As a result, the commons that were earlier ubiquitous in the form of village pastures, village ponds, threshing grounds, tanks, rivulets, river beds, etc. have now become increasingly uncommon. We need to acknowledge that the tragedy of the commons impacts the poor disproportionately as their dependence on the CPRs is the greatest. Livelihood policies and programmes need to actively prioritize strengthening local governance and conserving the commons. We opine that as the CPRs contribute significantly in supporting the livelihoods of the poor, they should be recognized as cardinal to the programmes directed towards reducing rural poverty and inequality.

Exercises

1. Ostrom challenged the conventional wisdom of Hardin's 'tragedy of the commons' and showed with empirical evidence that local communities can successfully manage their common pool resources. But how generalizable are Ostrom's principles of governing the commons when we increase the geographical scale from local to national to global? Discuss by referring to the following two publications.

   - Araral (2014) suggests that Ostrom's findings are more relevant for small, locally governed commons, while Hardin's theory holds true for large scale, national, regional, and global commons. Take the example of loss of tropical rainforests, air pollution in South Asia, depleting groundwater in large parts of India, pollution of rivers, global climate change, global biodiversity loss, etc.

     https://www.sciencedirect.com/science/article/pii/S1462901113001470#

   - Ostrom et al. (1999) suggest how conservation of global CPRs can take lessons from successful local management of the commons.

     http://citeseerx.ist.psu.edu/viewdoc/download?doi=10.1.1.473.891&rep=rep1&type=pdf
2. Familiarize yourself with Ostrom's eight design principles of CPRs. Do you agree that Ostrom's CPR framework provides an enriched view of natural resource management based livelihoods? Discuss in the context of the research work of Krause et al. (2019) where they study the future of primates in the forests of Nigeria.
   https://link.springer.com/content/pdf/10.1007/s11625-019-00667-y.pdf
3. Cox et al. (2010) analyse 91 research studies to evaluate Ostrom's eight design principles for community-based natural resource management. Discuss the reformulation of the design principles that they propose based on the findings of this analysis.
   https://www.ecologyandsociety.org/vol15/iss4/art38/main.html
4. The Scheduled Tribes and Other Traditional Forest Dwellers (Recognition of Forest Rights) Act, 2006 was enacted by the Indian parliament to undo the historical injustice meted out to the forest dwellers during colonial times when the community forests were nationalized labelling them as encroachers. Section 3(1)(i) of this Act under "Chapter 2 – Forest Rights" provides community forest resource management rights to the local community. Critically analyse the model of forest governance envisioned in this Act using the lens of Ostrom's eight design principles for successful CPR management.
   https://tribal.nic.in/FRA/data/FRARulesBook.pdf
5. The study by Persha et al. (2011) suggests that sustainable forest system outcomes measured in terms of species richness and subsistence livelihoods are far more likely in larger sized forests when local forest users participated in forest rulemaking, and the forests provisioned commercial livelihoods. Chhatre and Agrawal (2008) on the other hand found that better forest regeneration occurs in small to medium size forests which provide subsistence dependence, have low commercial value, with collective action, and local rule making. Compare and contrast these two studies in terms of their dependent variables and the causal linkage with the independent variables.
   https://www.iccaconsortium.org/wp-content/uploads/2015/08/example-local-rulemaking-persha-2011.pdf
   https://www.pnas.org/content/pnas/105/36/13286.full.pdf
6. Levine and Richmond (2015) examine two fishery co-management programs in the western Pacific region of the USA at Samoa and Hawaii. While these two regions share similar ecological and political contexts, their outcomes have been divergent. Examine these two co-management

programs using Ostrom's design principles and other factors to analyse what contributed to their success or failure.
https://doi.org/10.1016/j.marpol.2015.08.019

7. Typically research on the commons tends to construct three major policy alternatives to manage the commons, namely centralized state ownership, privatized management, and user self-governance as three mutually exclusive models. Sarker et al. (2015) apply Ostrom's design principles for managing coastal fisheries common in present-day Japan. They investigate the role of the state in enhancing user self-governance without taking ownership of the resource or participating in day-to-day management and governance. Study this paper and discuss the proactive role that the state can play to increase the utility of Ostrom's eight design principles and in polycentric governance.
https://doi.org/10.1016/j.ecolecon.2015.06.019
8. Is natural resource management by ecosystem people effective, or whether small population size and a lack of technology meant that exploitation levels were too low to have a significant impact upon the environment? Kahui and Richards (2014) evaluate the ecosystem management by the indigenous Maori in the south of New Zealand. Discuss the performance of the indigenous management systems when viewed through the lens of Ostrom's design principles. https://doi.org/10.1016/j.ecolecon.2014.03.006
9. Sarker and Itoh (2001) while analysing Japanese irrigation common pool resources (CPR) highlight the role of the state in building need-based large irrigation projects and entrusting them to the organization that the irrigators collectively form to manage their CPRs. Study this paper and discuss the proactive role that the state can play in a different context—'Management of Community Forest Resources by Indigenous People' to increase the utility of Ostrom's eight design principles and in polycentric governance.
https://doi.org/10.1016/S0378-3774(00)00125-6
10. Institution building of community-based natural resource management (CBNRM) organizations is challenging as often the poorest who depend on natural resources the most may not find adequate representation in the village-level institutions that are mandated to manage these resources. Also, planning needs to integrate at a higher geographical unit—typically at the landscape or watershed level. Discuss the learnings that IFAD has accumulated through its support to natural resource management (NRM) projects on how to resolve these institutional issues in CBNRM organizations in the 'how to do notes' series titled 'Strengthen Community-based Natural Resource Management Organizations'.
https://www.ifad.org/documents/38714170/40184028/How+to+strengthen+community-based+natural+resource+management+organizations.pdf/444e08c4-35b8-4150-b3a5-850f89ed5e04
11. Land tenure is a complex issue within pastoralism and IFAD has accumulated learnings of factors that lead to secure tenure or resulted in conflicts in the 'how to do notes' series titled, 'How to Prevent Landuse Conflicts in Pastoral Areas'. This note also discusses the frameworks, methods, and tools that can be used as a part of conflict analysis and project design. Discuss the learnings from this documentation on how to analyse and resolve pastoral conflicts.
https://www.ifad.org/documents/38714170/40184028/LandUseConflicts.pdf/4da68519-6c21-bc00-67df-d7e75aba9543
12. IFAD developed the Guidelines for 'Impact Evaluation of Land Tenure and Governance Interventions' in 2019. These guidelines discuss several frameworks, methods, and tools that can be used for this purpose such as the theory of change/logic model, regression discontinuity, difference in difference and matching. Build on the learnings from this documentation to develop an evaluation plan of a community forestry initiative.
https://www.ifad.org/documents/38714170/41206887/guidelines_ie_land.pdf/4435be07-829e-6d3d-444e-8f01403605d9

13. Theesfeld (2004) studies the constraints in managing the irrigation systems in Bulgaria using Ostrom's design principles for collective action. She reports that individualistic behaviour, distrust, envy, and pessimism towards collective action constrain the evolution of social capital. In this scenario, what institutional arrangement would you propose to manage the irrigation systems in Bulgaria? Give reasons in support of your argument.
    https://doi.org/10.1016/j.worlddev.2003.11.001
14. Morrow and Hull (1996) study the Yanesha forestry cooperative in Peru and show that indigenous CPR regimes that are not self-organized, in which a significant portion of the institutional rules is designed by donor agencies and external influences, are unlikely to meet Ostrom's eight design principles associated with durability. In such cases what should be a more effective strategy to develop sustainable self-governing institutions?
    https://doi.org/10.1016/0305-750X(96)00064-2
15. The Foundation for Ecological Security (FES) is an NGO working on reversing the 'tragedy of the commons' in India by restoring commons and building local institutional capacity on scale. Discuss the lessons learnt from restoring the commons and building institutional capacity in the western state of Rajasthan in India.
    http://fes.org.in/impact/external/innovations-in-common-land-development.pdf

Additional resources

1. Video on Promise of the Commons, 22 minutes, directed and produced by John D. Liu & Patrick Augenstein available at https://commonstransition.org/the-promise-of-the-commons/
2. Video – From Tragedy to Promise of Commons – South Asia Webinar Series, Global EverGreening Alliance, 1 h, 44 min, available at: https://www.youtube.com/watch?v=WH1O7AF7k1k
3. Video on Free Rider Problem by Vanya Bisht, 10 min, available at: https://www.youtube.com/watch?v=LTgiiDNyFkc
4. Video on Ostrom's Design Principles by Vanya Bisht, 7:45 min, available at:
   https://www.youtube.com/watch?v=QTQPy9tC5WE

References

Agrawal, A. (2001). Common property institutions and sustainable governance of resources. *World Development, 29*(10), 1649–1672.

Angelsen, A., Jagger, P., Babigumira, R., Belcher, B., Hogarth, N. J., Bauch, S., … Wunder, S. (2014). Environmental income and rural livelihoods: A global-comparative analysis. *World Development, 64*, S12–S28.

Araral, E. (2014). Ostrom, Hardin and the commons: A critical appreciation and a revisionist view. *Environmental Science & Policy, 36*, 11–23.

Chhatre, A., & Agrawal, A. (2008). Forest commons and local enforcement. *Proceedings of the National Academy of Sciences, 105*(36), 13286–13291.

Cox, M., Arnold, G., & Tomás, S. V. (2010). A review of design principles for community-based natural resource management. *Ecology and Society, 15*, 4.

Cousins, B. (2000). Tenure and common property resources in Africa. In C. Toulmin & J. F. Quan (Eds.), *Evolving land rights, policy and tenure in Africa* (pp. 151–180). DFID/IIED/NRI.

Frischmann, B. M. (2013). Two enduring lessons from Elinor Ostrom. *Journal of Institutional Economics, 2013*(9), 387–406.

Ghate, R., Mehra, D., & Nagendra, H. (2009). Local institutions as mediators of the impact of markets on non-timber forest product extraction in central India. *Environmental Conservation, 36*(1), 51–61.

Hardin, G. (1968). Tragedy of the commons. *Science, 162*, 1243–1248.

Jodha, N. S. (1986). Common property resources and rural poor in dry regions of India. *Economic and Political Weekly, 21*(27), 1169–1181.

Kahui, V., & Richards, A. C. (2014). Lessons from resource management by indigenous Maori in New Zealand: Governing the ecosystems as a commons. *Ecological Economics, 102*, 1–7.

Krause, T., Nielsen, T., Guia-Diaz, L., Lehsten, V., Olsson, O., & Zelli, F. (2019). What future for primates? Conservation struggles in the forests of Cross River State, Nigeria. *Sustainability Science, 14*(6), 1515–1529.

Leach, M., Mearns, R., & Scoones, I. (Eds.). (1997). Community based sustainable development: Consensus or conflict. *IDS Bulletin, 28*(4), 12.

Levine, A., & Richmond, L. (2015). Using common-pool resource design principles to assess the viability of community-based fisheries co-management systems in American Samoa and Hawai'i. *Marine Policy, 62*, 9–17.

Mehta, L., Leach, M., Scoones, I.. (2001). Governance in an Uncertain World. IDS Bulletin 32.4. Editorial: Environmental Institute of Development Studies.

Morrow, C. E., & Hull, R. W. (1996). Donor-initiated common pool resource institutions: The case of the Yanesha Forestry Cooperative. *World Development, 24*(10), 1641–1657.

Ostrom, E. (1990). *Governing the commons: The evolution of institutions for collective action*. Cambridge University Press.

Ostrom, E. (1998). A behavioral approach to the rational choice theory of collective action: Presidential address, American Political Science Association, 1997. *American Political Science Review, 92*(1), 1–22.

Ostrom, V., & Ostrom, E. (1977). Public goods and public choices. In E. S. Savas (Ed.), *Alternatives for delivering public services: Toward improved performance* (pp. 7–49). Westview Press.

Ostrom, E., Burger, J., Field, C. B., Norgaard, R. B., & Policansky, D. (1999). Revisiting the commons: local lessons, global challenges. *Science, 284*(5412), 278–282.

Pagdee, A., Kim, Y. S., & Daugherty, P. J. (2006). What makes community forest management successful: A meta-study from community forests throughout the world. *Society and Natural Resources, 19*(1), 33–52.

Persha, L., Agrawal, A., & Chhatre, A. (2011). Social and ecological synergy: Local rulemaking, forest livelihoods, and biodiversity conservation. *Science, 331*(6024), 1606–1608.

Persha, L., & Andersson, K. (2014). Elite capture risk and mitigation in decentralized forest governance regimes. *Global Environmental Change, 24*, 265–276.

Quinn, C. H., Huby, M., Kiwasila, H., & Lovett, J. C. (2007). Design principles and common pool resource management: An institutional approach to evaluating community management in semi-arid Tanzania. *Journal of Environmental Management, 84*(1), 100–113.

Sarker, A., Ikeda, T., Abe, T., & Inoue, K. (2015). Design principles for managing coastal fisheries commons in present-day Japan. *Ecological Economics, 117*, 32–38.

Sarker, A., & Itoh, T. (2001). Design principles in long-enduring institutions of Japanese irrigation common-pool resources. *Agricultural Water Management, 48*(2), 89–102.

SAPPLPP (2009). Innovations in common land development: Strengthening Institutional and physical spaces for poor livestock keepers. South Asia Pro Poor Livestock Policy Programme. Code: SAGP02, Good Practice Note, Delhi, India.

Tambe, S., Patnaik, S., Upadhyay, A. P., Edgaonkar, E., Singhal, R., Bisaria, J., Srivastava, P., Dahake, K., Hiralal, M. H., Tofa, D., Telharkar, S., Edlabadkar, V., Dethe, V., Sekhar K. (2019). "Co-production of sustainable bamboo management solutions in Eastern Maharashtra", Research study report, Indian Institute of Forest Management (IIFM), Bhopal, pp 181, ISBN: 978-81-943801-0-8.

Theesfeld, I. (2004). Constraints on collective action in a transitional economy: The case of Bulgaria's irrigation sector. *World Development, 32*(2), 251–271.

TISS (2018). What are the benefits of community forest rights? A tale of Pachgaon in Chandrapur district, Maharashtra, Tata Institute of Social Sciences, Mumbai, India.

Wade, R. (1988). *Village republics: Economic conditions for collective action in South India*. Cambridge University Press, Cambridge, UK.

Wunder, S. (2001). Poverty alleviation and tropical forests—What scope for synergies? *World Development, 29*(11), 1817–1833.

Chapter 6
Livelihood Trajectories

Abstract Livelihoods are inherently dynamic in response to shocks, stresses, and opportunities, resulting in households either escaping poverty, falling back, or sustaining their status. Livelihood trajectory research brings into focus concepts such as chronic and transient poverty, vulnerability and resilience, and the pathways into and out of poverty. It attempts to answer three key questions: how poverty is perceived, how it has changed, and what are the drivers of change. Mixed methods are used to answer these questions and capture the quantitative aspects in a transition matrix, and the qualitative aspects using life-history interviews. The diversity of these trajectories has been classified through a livelihood strategy typology as 'dropping out', 'hanging in', 'stepping up', and 'stepping out'. In diverse settings across the global south, livelihood trajectory studies indicate that while reasons for ascent are many and differ widely, the reasons for descent are often related to death or extended illness of the household head. These studies advocate that ending poverty will require two sets of interventions—one to accelerate ascent and the other to block descent. We opine that the livelihood trajectory approach has immense potential in informing livelihood projects and development policy which has not been fully realized.

Keywords Poverty dynamics · Chronic poor · Transient poor · Mixed methods · Panel dataset · Stages-of-progress · Longitudinal study · Transition matrix · Life stories · Explanatory sequential · Socio-economic mobility

6.1 The Dynamic Nature of Poverty

Poverty is inherently dynamic, with households escaping and falling back into poverty simultaneously (Krishna, 2007). How do we then differentiate between the chronic poor and the transient poor? Hulme and Shepherd (2003) define the chronic poor as those who are poor for much of their lifespan and may transmit their poverty to the next generations. The chronic poor comprise the *always poor* whose poverty score is always below a defined poverty line and the *usually poor* who may not always be below this poverty line but their mean poverty score is less than this poverty line (Fig. 6.1). The transient poor are the *occasionally poor* whose mean poverty score is above the poverty line but have spent some time below it, and the *never poor* who are always above the poverty line. Figure 6.1. Poverty depth or poverty severity is the distance the household's poverty status lies below the poverty line. The more the poverty severity, the more will be the effort needed to escape poverty Fig. 6.2. Livelihood trajectories are adversely impacted by shocks and stresses, while opportunities enable upward mobility. Shocks are sudden sharp events that harm wellbeing such as illness, loss of job, disasters, crime, accidents, market dynamics, pest attacks, etc., while stresses are long-term trends or pressures such as soil degradation, social discrimination, climate variability, etc. (Sagara, 2018). These shocks and stresses impact livelihood trajectories of vulnerable households especially when they lack resilience.

S. Tambe, *Teaching and Learning Rural Livelihoods*, Sustainable Development Goals Series,
https://doi.org/10.1007/978-3-030-90491-3_6

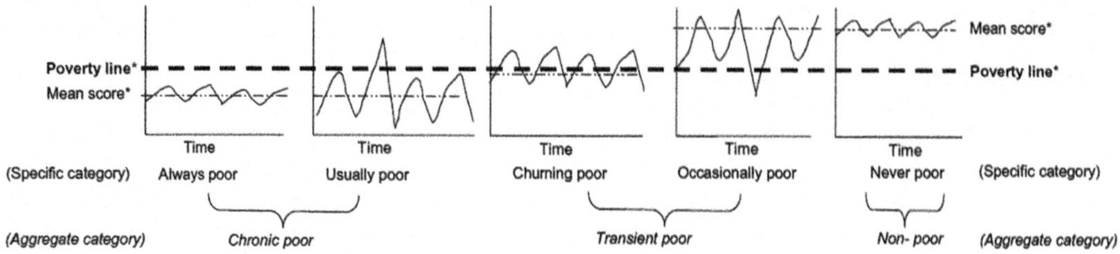

Fig. 6.1 Poverty dynamics represented by the chronic poor, transient poor, and the non- poor (Adapted from Hulme & Shepherd, 2003)

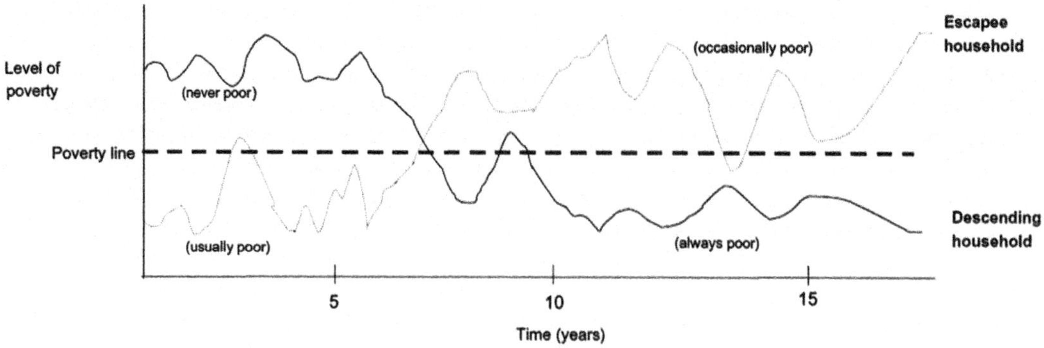

Fig. 6.2 Poverty dynamics: escaping and descending households (Adapted from Hulme & Shepherd, 2003)

6.2 Livelihood Trajectories

Livelihood trajectory refers to the path that households or groups take over time while also describing and explaining the direction and pattern (Bagchi et al., 1998). Assessment of the socio-economic mobility of households is gained through temporal, snapshot surveys, and life-history stories that trace these household trajectories. As it is not possible to continuously monitor the changes in a person's life, we tend to look at aggregate assessments by taking snapshots at particular intervals of time typically in multiples of 5 years. Analysis of common life trajectory patterns (Fig. 6.3) reveals that while improvements happen slowly over a longer period, these are interspersed with rapid declines due to shocks, resulting in a saw-tooth livelihood trajectory pattern (Davis & Baulch, 2011). The overall livelihood trajectory following a shock depends on the severity of the shock and the vulnerability and the resilience of the household.

These livelihood trajectories are classified by various authors based upon the change in poverty status or the livelihood strategy adopted (Table 6.1). Building on poverty dynamics these trajectories are categorized as 'remained poor', 'escaped poverty', 'became poor', and 'remained non-poor' (Krishna, 2006). While based on the livelihood strategy they are classified into four types. The first is 'hanging in' wherein the face of adversity the households struggle to maintain their livelihood levels. The second is 'stepping up' or significantly enhancing their existing livelihood activities. The third is 'stepping out' where the accumulated assets are used by the households to diversify into new activities. The fourth is 'dropping out' which is a declining status often due to chronic poverty or ill health (Dorward et al., 2009; Mushongah, 2009). Sallu et al., 2010 categorized households as 'accumulator', 'diversifier', and 'dependent' based on their livelihood strategy. The 'accumulators' have a tendency to specialize like accumulating large number of livestock by the village elite. The 'diversifiers' had a more varied strategy and pursued multiple livelihood options. The 'dependents' lacked livelihood assets and were dependent on social safety nets. Scoones et al., (2012) identified a 'middle farmer' group, reliant on 'accumulation from below' through petty commodity production, existing alongside other peasants whose livelihoods remain vulnerable, with prospects for accumulation currently limited. While there are others who are

Fig. 6.3 Conceptualizing the common life trajectory patterns (Source: Davis & Baulch, 2011)

Direction	Pattern	Depiction	Number of Cases	Weighted Percent of Cases
Stable	Smooth		8	1.47
Improving	Smooth		3	1.43
Declining	Smooth		2	0.36
Stable	Saw-tooth		135	44.98
Improving	Saw-tooth		76	26.15
Declining	Saw-tooth		30	6.90
Declining	Single-step		2	0.48
Declining	Multi-step		37	18.22
Total			293	100

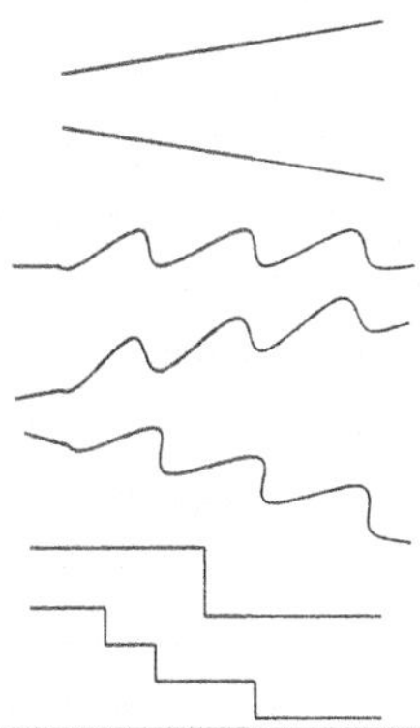

Table 6.1 Classification of household livelihood trajectories by various scholars

Author	Livelihood trajectories
Krishna (2006)	'Remained poor', 'escaped poverty', 'became poor', and 'remained non-poor'
Dorward et al. (2009), Mushongah (2009)	'Hanging in', 'stepping up', 'stepping out' and 'dropping out'
Sallu et al. (2010)	'Accumulator', 'diversifier' and 'dependent'
Scoones et al. (2012)	'Accumulation from below' and 'accumulation from above'
Pritchard et al. (2017)	'Hanging in', 'stepping up', 'stepping out' and 'dropping out, going backward or muddling through'
Thanh et al. (2021)	'Accumulating', 'fluctuating' and 'marginalizing'

'accumulating from above', through patronage and corruption. This group was socio-economically influential due to these power networks. Other than assessing household-level mobility, this approach can also be applied to assess socio-economic mobility of specific 'target groups' such as prawn farmers, forest dwellers, landless households, women-headed households, etc. (Bagchi et al., 1998).

6.3 Assessing the Trajectory Using Mixed Methods

Understanding poverty dynamics necessitates the use of mixed methods that supplement, complement, and counter one another (Kothari & Hulme, 2004). Creswell (2014) identified three basic mixed methods, namely convergent parallel, explanatory sequential, and exploratory sequential. In the livelihood trajectory approach, a combination of exploratory sequential and explanatory sequential methods is used. In exploratory sequential, first, the local perception of poverty is assessed qualitatively followed by quantification of the livelihood trajectories into categories in a transition matrix. Then the explanatory sequential method is used wherein qualitative household interviews are conducted category-wise to explain the factors influencing the pathways into and out of poverty. Kothari and Hulme (2004) point out that the livelihood trajectory approach highlights not only 'what' has changed, but also provides insights on the 'why' and 'how', and thus results in a deeper understanding of the lived realities of the poor. To assess the poverty status, we discuss three methods—the stages-of-progress, participatory wellbeing ranking (PWR), and household expenditure surveys.

6.3.1 Stages-of-Progress Approach

Krishna (2006) conceptualized the stages-of-progress (SOP) approach to assess the livelihood trajectories of households. This approach involves seven sequential steps to generate the transition matrix and life stories (Box 6.1).

Box 6.1 Stages-of-Progress Approach
Krishna (2006) proposed the 'stages of progress' approach involving the following seven steps:

Step 1. Assembling a diverse and representative community group.
Step 2. Presenting the objectives of the exercise.
Step 3. Defining collectively what it means for a household to be regarded as poor and ascertaining the steps (stages) it follows to climb out of poverty.
Step 4. Inquiring households' present poverty status and 20 years back.
Step 5. Assigning households to particular categories.
Step 6. Inquiring about reasons for escape and descent in respect of a random sample of households.
Step 7. Following up by interviewing household members.

Source: Krishna (2006).

In this approach, the stages that households cross while escaping poverty are ascertained using participatory methods with a representative group. What are the expenditures a household incurs as it climbs out of poverty? In south India, the stages that were identified commonly across villages were food security, minor house repair, debt repayment, buying clothing to wear outside, digging irrigation well, constructing a brick house, purchasing cattle, purchasing a bicycle, buying gold, purchasing a motorcycle, purchasing land, purchasing a tractor, and buying a television (Krishna, 2006). These first four stages comprise the local perception of poverty. Once the households escaped the fourth stage, i.e. clothing to wear outside, they were regarded as non-poor. Following this, a complete list of all the households in the village is prepared and the poverty status of each household is ascertained both for the present and in the past (i.e. 25 years back). Based on this assessment across two snapshots, the livelihood transition matrix is generated by assigning the households to one of the four categories, namely 'remained poor' (PP), 'escaped poverty' (PN), 'became poor' (NP), and 'remained non poor' (NN). PP stands for poor both in the past and present, PN is poor in the past but not poor in the present, NP is non-poor in the past but poor in the present, while NN is not poor both in the past and the present (Table 6.2).

In this example, 20% of the households remained poor (PP), 40% escaped poverty (PN), 10% became poor (NP), while 30% (NN) remained non-poor. So the net impact on poverty is the net impact of households who escaped poverty minus those that fell into poverty (PN – NP), i.e. 60% – 30% = 30%. Following this, a few households are randomly selected from each of the four categories, and their life stories are documented to ascertain the drivers of change. Krishna (2006) found health related expenses, marriage and funerals, drought, and large family size as the main reasons for the descent into poverty, with livelihood diversification and irrigation being the main reasons for escape. This approach helps to ascertain not only *who* is poor, but also *why* they are poor, and *what* are the factors influencing the escape or descent into poverty.

Steps 3 and 4 which involve assessing the household poverty status in the present and 25 years ago can be replaced with two alternative methods, namely participatory wellbeing ranking (PWR) and household expenditure surveys as described below, with the other steps remaining the same.

Table 6.2 Livelihood transition matrix over a 20-year timespan

	20 years back	
Present	Poor	Non-poor
Poor	20% Remained poor (PP)	10% Became poor (NP)
Non-poor	40% Escaped poverty (PN)	30% Remained non-poor (NN)
Poor now	PP + NP = 20% + 10% = 30%	
Poor 20 years back	PP + PN = 20% + 40% = 60%	
Net poverty status	60% - 30% = 30%	

Table 6.3 Criteria for participatory wellbeing ranking (PWR)

Parameter	Very poor	Poor	Middle	Rich
Monthly income (INR)	<5000	5000–1000	10,000–25,000	>25,000
Land ownership	Landless	Landless	<5 acres	>5 acres
Livestock ownership	1–3 goats	1–2 cows	2–5 cows	>5 cows
Education of head	Primary	Secondary	Senior secondary	Graduate
House type	A hut made from stone, mud, and thatch	Brick house with a tin roof	Brick house with concrete roof	A big, concrete, well-furnished house
Cooking fuel	Firewood and cow dung	Firewood and cow dung	LPG cylinder	LPG cylinder
Vehicle ownership	Bicycle	Motorcycle	Motorcycle	Car
Main source of income	Wage labour	Skilled labour	Small business	Big business or salaried job

6.3.2 *Participatory Wellbeing Ranking*

In this tool, a focus group discussion with a representative group is held to first develop a criterion which is then used to segregate the households into categories. This criterion generally comprises parameters such as land ownership, livestock ownership, vehicle ownership, house type, main livelihood, education, etc. It may vary from village to village as we are looking at poverty from the eyes of the villagers (Table 6.3). The purpose of this participatory exercise is to segregate the households into wellbeing categories both for the present and 20 years back. The transition matrix is then generated with its four livelihood trajectory categories followed by steps 6–7 that focus on the qualitative aspects (Boxes 6.2, and 6.3, Fig. 6.4).

Box 6.2 Livelihood History Interview with Angel

Angel was brought up by her mother and grandparents in a rural community. She had her first child when she was 12, but continued going to school till Form 2. She had to drop out because her family ran out of money. To earn a living, she left her baby with her mother and at the age of 14 migrated to town, and took up work as a house girl, leaving her two-year-old son behind. But her employer exploited her: she received no wages, and after a year she quit. During this time her mother had died, and her daughter had been placed in the care of the child's father. Instead of returning 'home', Angel moved to the informal settlement at Plot Shumba, a privately owned peri-urban site near a large town in Midlands Province in Zimbabwe, and eked out a living by vending. In February 2003, as part of the state's drive against informality, all 50 or so homes in the plot were destroyed. Angel was severely beaten and slept rough at a bus shelter for a month. After the landowner of the plot obtained a court ruling, Angel was able to return and rebuild her house and life. In the next two years, Angel's fortunes improved; with the help of a cousin, she found a job at a bar in a local gold mine and fell in love with a gold panner. However, when the mine closed, both lost their jobs. Angel returned to informal trading, and her boyfriend turned his hand to informal mining. In March 2005, she gave birth to their first son. But soon afterward her boyfriend died, when the pit he was mining in collapsed. In 2006, the state razed Plot Shumba to the ground as part of Operation Murambatsvina. Angel, 25, diagnosed HIV positive now lives alone with her 19-month old son, and with her health failing, is reliant on neighbours and a local NGO to provide food for her and her son.

Source: Bird and Prowse (2008), Addison et al. (2008).

Box 6.3 Assessing Poverty Pathways: From Rain-Fed to Organic Vegetable Farming

In a forest fringe village in central India most of the households engaged in subsistence rain-fed farming and cultivated rice, maize, and wheat in their upland fields. Their fertile lands were submerged by the backwaters of a dam for which they were compensated. These backwaters would recede during the winters, exposing fertile silt-rich lands locally known as *kachaar*. The *kachaar* lands are very productive and served as village grazing lands. In 2011 a public micro-credit programme entered the village with a street play, highlighting how the sickness of children and the subsequent medical treatment resulted in indebtedness. After some initial doubts, the women agreed to organize themselves into five self-help groups (SHGs) and to federate at the village level. They planned to take up vegetable farming in the *kachaar* lands once the floodwaters receded. The main hurdles in vegetable farming were a lack of machinery for lift irrigation from the dam and controlling the free-ranging cattle. The SHG members convinced the villager elders to regulate the cattle and provisioned loans to the members for the procurement of pumps and pipes. Around 2–3 acres of *kachaar* land was made available to each of the 62 households. These fertile lands produced a bumper harvest of bitter gourd (*karela*). Soon their gourd became popular as it was organic and had a longer shelf life, enabling transport to nearby towns. Organic vegetable farming became their cash crop with a readymade market in the wholesale market providing additional household incomes of Rs 30–40,000 during these 3–4 months.

In 2016, a poverty dynamics study was taken up in this village. The participants profiled the very poor as those who were landless, lived in a thatched house, did not own livestock, had no farming income, and labour work was the sole income source. The poor were identified as those who possessed up to two acres of land, lived in a temporary house having a tiled roof, and

(continued)

Box 6.3 (continued)

owned a few livestock. They had two sources of income—small scale farming and labour work. The middle class was the rest who owned two to five acres of land which was irrigated and possessed livestock as well. They had three or more sources of income, namely farming, owned shops, wage employment, etc. On constructing the transition matrix, the study found that due to the intervention of the micro-credit programme over five years, 30 (48%) poor households had escaped poverty and graduated to middle class.

Poverty category	Households	
	2011	2016
Very poor	24	7
Poor	34	21
Middle class	4	34
Total	**62**	**62**

Household interviews revealed that the pathways out of poverty were vegetable farming, livestock rearing, educated children taking up jobs, women taking up village-level jobs, and income generation activities such as flour mill, grocery store, etc. The households stagnating in poverty were the ones with young children needing constant nursing, thereby preventing the mothers from actively participating in livelihood activities. Also, there were a few vulnerable women, who had been deserted by their husbands and a few widows who were staying with their parents who would need special attention.

Source: Patnaik et al. 2017.

6.3.3 Household Expenditure Surveys

In this method, longitudinal household expenditure surveys taken up 20 years apart form the basis to construct the transition matrix. The households are categorized into four categories as described in Table 6.2. This approach is time-taking and covering 200 households at the rate of 10 households per day would take us nearly a month. In this duration, the PWR approach can cover 5–6 villages in a more comprehensive and in-depth manner.

6.4 Qualitative Assessment of the Pathways

While the study of livelihood trajectories has been dominated by panel datasets, qualitative methods can complement these studies by explaining the poverty dynamics (Kothari & Hulme, 2004). Steps 6–7 (Box 6.1) take the form of an explanatory sequential mixed method where following the quantitative poverty assessment, qualitative household interviews help explain these trajectories. A certain percentage of households from each of the four categories are randomly selected to document their life history. These life-history stories so generated help in deepening our understanding as to why certain households fall into poverty while others sustain their status (Kothari & Hulme, 2004). The question for the upwardly mobile households is what enabled the jump, and for the ones that descended, what made them vulnerable to the shock (Box 6.2).

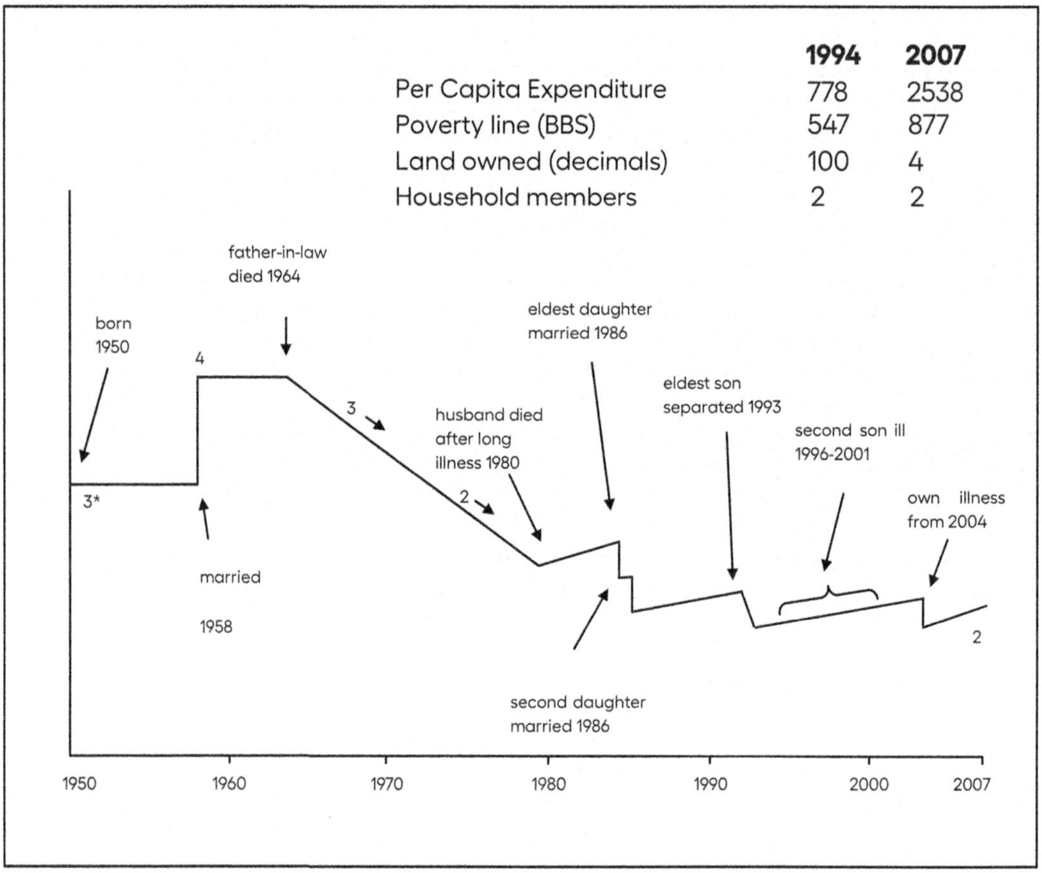

Fig. 6.4 Expenditure as an imperfect indicator of wealth—quantitative versus qualitative poverty pathways (Source: Davis & Baulch, 2011)

6.5 Reconciling Quantitative and Qualitative Results

How does one integrate quantitative and qualitative data in a poverty dynamics study? Davis and Baulch (2011) carried out qualitative household interviews and compared them with a larger quantitative panel survey of household data to assess the life trajectories in rural Bangladesh. They analyse the transition matrix, poverty pathways, and factors influencing poverty dynamics and find that livelihood trajectories derived from quantitative household expenditure surveys may not match with the qualitative life trajectories generated from household life stories (Fig. 6.4). In this case, while quantitative methods of per capita expenditure indicate that the household is improving, the qualitative methods uncover a household struggling with chronic illness and meeting its health expenses from the distress sale of land that declined from 100 to 4 decimals. This comparison underscores that expenditure alone is an imperfect indicator of poverty status and needs to be coupled with other indicators such as ownership of land, livestock, etc. to give a realistic picture.

6.6 Results of Livelihood Trajectory Studies

Here we share the results of livelihood trajectory studies spanning the global south. Krishna (2007) took up a series of studies covering several countries of Africa, Asia, and Latin America. He found that while multiple factors are linked to ascents, ill health and high healthcare costs overwhelmingly constitute the most important reason for households to descend into poverty. Ill health not only reduced the earning capacity of the households but also added considerably to the expenditure, thereby striking a double blow. Social and customary expenses on marriages and funerals were another major cause of descent. Income diversification through a new source of income from on-farm and off-farm strategies was an important pathway out of poverty. On-farm strategies included pursuing new crops, new techniques, cash crops, livestock, etc. while off-farm strategies included petty trade, small businesses, wage employment, etc.

The stages-of-progress till the poverty cut-off in rural Kenya included six stages, namely food, clothing, house repair, primary education, starting small business, and purchasing small livestock (Kristjanson et al., 2010). Also, the pathways differed across the agro-ecological zones such as fertile lands, marginal lands, agro-pastoral, pastoral, and urban areas as the context varied. In rural Zimbabwe the criteria for wealth included the size of land and cattle holding, farm production, remittances, etc. (Mushongah & Scoones, 2012). The pathways into poverty were due to death or extended illness of the household head. Medical expenses were a significant burden and deaths also entailed elaborate funerals. Loss of critical assets such as livestock to droughts and disease was also linked to a decline in wealth status. An increase in rank was associated with the accumulation of assets such as cattle, land, and farm equipment. The ability to withstand shocks gets enhanced by asset accumulation such as cattle and cash remittances. A study by Pritchard et al. (2017) in north India points to the vital importance of land ownership as the major material asset and a source of social status. Wilson (2004) in a study on understanding the dynamics of socio-economic mobility in central India concluded that while the rural economy had experienced considerable structural changes, it was mostly the historically advantaged households in terms of social and financial capital that were able to experience considerable improvements in their wellbeing. For the economically disadvantaged and the lower caste households, the typical strategy to move out of poverty was urban migration and gradual accumulation of resources by drawing on their experiences from the urban areas to improve their situation within the village.

6.7 Learnings from Practice

The livelihood trajectory approach provides an insight into the richness and complexity of household dynamics and how it affects asset accumulation. The beauty of this approach is that one can put a finger on the pulse of a site right away and intimately experience their struggles and strife. Implementing the livelihood trajectory approach can face a few practical challenges as well, and we share insights on how to address them.

6.7.1 Assessment of Poverty Status

For poverty assessment, the PWR is widely used in projects compared to the stages-of-progress (step 2) and household expenditure surveys, as it is easier to use and the past status can be assessed using recall methods. To enable comparison across villages in the project area, a standardized PWR criterion can function as a common benchmark. While implementing PWR, initially the local community may find it difficult to come up with a criterion to segregate the households into socio-economic categories. They may initially profess that all the households are similar and belong to one category. In such cases, additional probing is needed with the facilitator enquiring about the asset profile (land, livestock, house, child education, livelihood, etc.) of the poorest household in the village, and repeating it for the wealthiest household in the village. This assessment of the socio-economic extremes helps to create a wellbeing range within which the other categories can then be fitted.

6.7.2 Discerning the Internal and External Drivers

While documenting life stories, the drivers can be either intrinsic or extrinsic to the household. Intrinsic factors such as a household with small children prevent the mother from participating in income generating activities away from home especially if no elders are around to take care of children. Also, the availability of young healthy males in a household who can contribute their labour and avail non-farm employment opportunities is a big positive for the household. Similarly, external enablers can play a role in accelerating pathways out of poverty such as access to forest produce from nearby forests, connectivity to an industrial area where they can get employment, newly constructed village canal enabling winter irrigation, skilling through tailoring using sewing machines, village road aiding in connectivity with urban labour markets, access to a reservoir and pumps enabling lift irrigation, etc.

6.7.3 Limitations of the Process

The livelihood trajectory approach suffers from some limitations as well. Subjective assessment of wellbeing and the resultant criterion to delineate poverty categories may not be comparable across areas with diverse contexts (Bagchi et al., 1998). Also, as a part of the data collection is from a distant timeframe of 20 years, recall surveys may get impacted by information decay with time (Bagchi et al., 1998). Organizing a representative village gathering for the focus group discussion, covering all the hamlets, socio-economic groups, and age classes with balanced gender representation will need advanced preparatory work in the village. One needs to be cautious before scaling up the pathways out of poverty as the nature of the value chain to accommodate fresh entrants is not known.

6.8 Implications for Policy

Livelihood trajectories provide a realistic assessment of the past and present to plan for the future. These studies help in ascertaining the impact of shocks and stresses on livelihoods and the steps needed to strengthen them (Sallu et al., 2010). They provide guidance on the livelihood strategy focussing on two types of interventions, one to prevent descent into poverty and the other to promote ascent (Krishna, 2007). Hence, formulating better poverty reduction strategies will require two different sets of policy responses—one to block descent and the other to accelerate ascent. Consider the example provided in Table 6.4, where the transition matrix of a village having 100 households is provided. Here we observe that Village A where households are stagnating in poverty will need an expansion of livelihood promotion interventions, whereas in Village B where non-poor households descended into poverty will need to relook at their livelihood protection programmes.

Table 6.4 Livelihood transition matrix

Livelihood trajectories	Village A	Village B
Remained poor	80	5
Escaped poverty	10	10
Remained non-poor	5	5
Descended into poverty	5	80

The livelihood trajectory approach helps in collecting a rich dataset at the household level in a focus group discussion mode thereby saving time (3–4 h). Also, as this data is collected publicly, it has the benefit of oversight as the responses can be verified by the community members. While the accuracy of the data collected in household surveys relies solely on faithful responses from the respondent. The livelihood trajectory approach can thus inform project planning as one can ascertain what is working and what is not in a given village. This approach is especially useful to ascertain if the safety nets are preventing the household from falling into poverty and whether the livelihood ladders are assisting the households to escape poverty. These stories put a human face behind the numbers in the table or graph and thereby help the reader to relate to them. It also has the potential to influence policymakers as these stories are closer to truth without any statistical complexity and can be verified easily. This approach can be used in monitoring and evaluation as well when the wellbeing of the households has to be tracked and its attribution to the project interventions ascertained.

Exercises

1. The transition matrix from a livelihood trajectory study in four villages is tabulated below. It depicts how people have escaped or fallen back into poverty over the last 20 years expressed as a percentage. Answer True/False for questions a to d based on this.

Livelihood trajectories	Village A	Village B	Village C	Village D
Remained poor	80	5	5	5
Escaped poverty	10	80	10	5
Remained non-poor	5	10	80	10
Descended into poverty	5	5	5	80

   (a) Livelihood ladders in Village A are adequate.
   (b) Village C has effective safety nets.
   (c) Safety nets in Village D are adequate.
   (d) Village B and Village C performed the best.

2. Davis and Baulch (2011) supplement a large quantitative panel survey of households with life-history interviews to assess the socio-economic mobility and life trajectories in rural Bangladesh.

   - What are the difficulties associated with reconciling quantitative and qualitative data and how did they address it?
   - Compare the 'mixed methods' approach of Davis and Baulch (2011) with the 'stages-of-progress' method of Krishna (2006) to analyse poverty dynamics.

   https://www.trentu.ca/ids/sites/trentu.ca.ids/files/documents/Q2_WP42_BaulchandDavis.pdf
3. What is chronic poverty, who is chronically poor, why are people chronically poor, and how does this impact the design of the poverty reduction strategy? Hulme and Shepherd (2003) provide a useful analysis to address these questions. How does the categorization of the poor into chronic, transient, and non-poor provided in Fig. 6.1 of this study compare with the six categories of the BRAC graduation approach, namely destitute, extreme poor, moderate poor, vulnerable non-poor,

non-poor, and the wealthy (De Montesquiou et al., 2014, 14)? Discuss how policy choices are affected by looking at the poor not as a homogenous group, but as belonging to different categories?

https://www.sciencedirect.com/science/article/abs/pii/S0305750X0200222X

https://www.cgap.org/sites/default/files/graduation_guide_final.pdf

4. How does the poverty reduction strategy vary for the transitory and chronic poor? Barrett (2005) discusses the implications of this distinction on developmental policy with the construct of safety and cargo nets. The need to differentiate and target these groups separately with not only concerns of whom to assist, where, and when but also how and what. Discuss the various targeting strategies and developmental interventions you would design acknowledging this differentiation.

   https://ageconsearch.umn.edu/record/245921/files/Conf25th2003_4.pdf

5. Escaping poverty and becoming poor have different sets of reasons. Krishna et al. (2004) studied the poverty dynamics in 20 villages of Kenya and compared it with the findings of a similar inquiry in 35 villages of Rajasthan. Discuss how the policy implications of this study can help in the better design of poverty reduction programmes?

   https://cgspace.cgiar.org/bitstream/handle/10568/749/escaping%20poverty%2020.pdf?sequence=2

6. Livelihood aspirations and strategies of the poor are diverse. Dorward et al. (2009) developed three broad classifications of livelihood strategy, namely—'hanging in', 'stepping up', and 'stepping out'. Discuss how this approach can be used to inform practice especially in devising livelihood strategies.

   https://www.prismaweb.org/nl/wp-content/uploads/2017/06/Hanging-in-stepping-up-and-stepping-out-livelihood-aspirations-and-strategies-of-the-poor%E2%94%82Dorward-Anderson-Nava-Bernal-S%C3%A1nchez-Vera-Rushton-Pattison-Paz%E2%94%82April-2009.pdf

7. Setting up a longitudinal livelihood change study is challenging as the original households have to be traced, wealth perceptions evolve, and the impacts of a range of macro and local shocks and stresses have to be factored in. Mushongah and Scoones (2012) study livelihood change in rural Zimbabwe two decades apart (1986–87 and 2006–07). Discuss the use of Dorward et al. (2009) livelihood change framework, qualitative household interviews, and wealth ranking framework to unravel the intricacies of livelihood dynamics over a 20-year period.

   https://www.researchgate.net/profile/Ian_Scoones/publication/254242850_Livelihood_Change_in_Rural_Zimbabwe_over_20_Years/links/561a9f1c08ae044edbb20c2b.pdf

8. The dynamics of socio-economic mobility of two Indian villages is studied by Wilson (2004) using a comparative approach. Discuss the livelihood trajectories in these two villages with a special emphasis on the pathways into and out of poverty and the drivers of the same.

   https://www.odi.org/sites/odi.org.uk/files/odi-assets/publications-opinion-files/2474.pdf

9. The students will need to carry out 4–5 h of participatory fieldwork in a village for this exercise. The study needs to be preceded by an exploratory visit to the village where the consent of the villagers and suitable date and time are fixed in advance. The objective of the visit needs to be conveyed, indicating that it is a student initiative, to not raise any undue expectations. Ideally, a group of five students can be assigned to a village (or a smaller unit) comprising about 50–70 households. The village can be informed that a group of 15–20 participants who are representative of this village unit is expected for this exercise. The students need to submit a short report of 5000 words covering the study title, study area, research questions, methods used, findings, discussions, conclusions along with process documentation.

Research Questions

The study attempts to answer the following research questions:

- How does the local community perceive poverty? i.e. participatory wealth ranking criteria.
- What is the poverty status of the households? i.e. segregating them into wealth classes.
- What is the pattern of poverty dynamics? i.e. the percentage of households who have escaped poverty, fallen into poverty, remained poor, and remained non-poor.
- What are the poverty pathways in the village? i.e. factors driving escape or descent.

Methodology

Use a mix of quantitative and qualitative methods. In the exploratory part ascertain how the villagers define poverty classes and perceive the wealth class of the households. Then using explanatory methods, ascertain the reasons behind ascent and descent.

- Study area: Write an introduction to the village covering its location, physical features, google map, demography, social structure, livelihood aspects, infrastructure, etc.
- Sampling: Indicate the total households in the village, how many you covered for poverty assessment. Of the total households in the four wealth classes, how did you select the ones to be interviewed.
- Process documentation—Activity photos with captions.

Expected Learning Outcomes

- Experiential learning about poverty, village heterogeneity, and poverty pathways.
- Introduction to social sciences research— research questions, frameworks, methods, findings, discussions.

Additional Resources

1. Video of Prof. Anirudh Krishna on why people become poor.
   https://www.youtube.com/watch?v=PxcZnfXY2CA
2. Participatory wellbeing ranking process.
   https://meas.illinois.edu/wp-content/uploads/2017/02/MEAS-Participatory-Methods-Tip-Sheet-Wellbeing-Ranking-Analysis-1.pdf

6.8.1 References

Addison, T., Harper, C., Prowse, M., Shepherd, A., Barrientos, A., Braunholtz-Speight, T., ... & Moore, K. (2008). The Chronic Poverty Report 2008–09: Escaping Poverty Traps. *Manchester: Chronic Poverty Research Centre, Brooks World Poverty Institute*. Retrieved on 9th June 2021 at 20:00 from https://odi.org/en/publications/chronic-poverty-report-2008-09-escaping-poverty-traps/

Bagchi, D. K., Blaikie, P., Cameron, J., Chattopadhyay, M., Gyawali, N., & Seddon, D. (1998). Conceptual and methodological challenges in the study of livelihood trajectories: Case-studies in eastern India and Western Nepal. *Journal of International Development, 10*(4), 453–468.

Barrett, C. B. (2005). Rural poverty dynamics: Development policy implications. *Agricultural Economics, 32*, 45–60.

Bird, K. and Prowse, M. (2008). 'Vulnerability, Poverty and Coping in Zimbabwe'. UNU-WIDER Research Paper No. 2008/41. Helsinki, Finland: UNU World Institute for Development Economics Research. Retrieved on 10th June 2021 at 8:00 from http://www.wider.unu.edu/publications/working-papers/research-papers/

Creswell, J. W. (2014). *A concise introduction to mixed methods research*. SAGE.

Davis, P., & Baulch, B. (2011). Parallel realities: Exploring poverty dynamics using mixed methods in rural Bangladesh. *The Journal of Development Studies, 47*(1), 118–142.
De Montesquiou, A., Sheldon, T., Degiovanni, F., & Hashemi, S. (2014). From extreme poverty to sustainable livelihoods: A technical guide to the graduation approach. *CGAP and Ford Foundation*.
Dorward, A., Anderson, S., Bernal, Y. N., Vera, E. S., Rushton, J., Pattison, J., & Paz, R. (2009). Hanging in, stepping up and stepping out: Livelihood aspirations and strategies of the poor. *Development in Practice, 19*(2), 240–247.
Hulme, D., & Shepherd, A. (2003). Conceptualizing chronic poverty. *World Development, 31*(3), 403–423.
Kothari, U., & Hulme, D. (2004). *Narratives, stories and tales: Understanding poverty dynamics through life histories. GPRG-WPS-011* (p. 41). ESRC Global Policy Research Group.
Krishna, A., Kristjanson, P., Radeny, M., & Nindo, W. (2004). Escaping poverty and becoming poor in 20 Kenyan villages. *Journal of Human Development, 5*(2), 211–226.
Krishna, A. (2006). Examining pathways out of and into poverty in 36 villages of Andhra Pradesh. *World Development, 34*(2), 271–288.
Krishna, A. (2007). For reducing poverty faster: Target reasons before people. *World Development, 35*(11), 1947–1960.
Kristjanson, P., Mango, N., Krishna, A., Radeny, M., & Johnson, N. (2010). Understanding poverty dynamics in Kenya. *Journal of International Development, 22*(7), 978–996.
Mushongah, J. (2009) Rethinking vulnerability: Livelihood change in southern Zimbabwe, 1986–2006. Ph.D. dissertation, University of Sussex.
Mushongah, J., & Scoones, I. (2012). Livelihood change in rural Zimbabwe over 20 years. *Journal of Development Studies, 48*(9), 1241–1257.
Patnaik, S. Tambe, S., Bisaria, J., Pandey, A., Sarma, U. K., & Ananda Vadivelu G. (2017). *Evaluation of Madhya Pradesh Deendayal Antodaya Yojana – State Rural Livelihoods Mission (MPDAY-SRLM)*. Study Report, Indian Institute of Forest Management (IIFM), Bhopal, 2017, pp 205, ISBN: 978-81-943801-1-5
Pritchard, B., Vicol, M., & Jones, R. (2017). How does the ownership of land affect household livelihood pathways under conditions of deagrarianization? 'Hanging in', 'stepping up' and 'stepping out' in two north Indian villages. *Singapore Journal of Tropical Geography, 38*(1), 41–57.
Sagara, B. (2018). *Resilience Measurement Practical Guidance Note Series 2: Measuring Shocks and Stresses*. Produced by Mercy Corps as part of the Resilience Evaluation, Analysis and Learning (REAL) Associate Award. Retrieved on 13th February 2021 at 11:00 from https://reliefweb.int/sites/reliefweb.int/files/resources/GN02_ShocksandStressesMsmt_Final.pdf
Sallu, S. M., Twyman, C., & Stringer, L. C. (2010). Resilient or vulnerable livelihoods? Assessing livelihood dynamics and trajectories in rural Botswana. *Ecology and Society, 15*, 4.
Scoones, I., Marongwe, N., Mavedzenge, B., Murimbarimba, F., Mahenehene, J., & Sukume, C. (2012). Livelihoods after land reform in Zimbabwe: Understanding processes of rural differentiation. *Journal of Agrarian Change, 12*, 503–527.
Thanh, H. T., Tschakert, P., & Hipsey, M. R. (2021). Moving up or going under? Differential livelihood trajectories in coastal communities in Vietnam. *World Development, 138*, 105219.
Wilson, C. (2004). Understanding the dynamics of socio-economic mobility: Tales from two Indian villages. Working Paper 236, Overseas Development Institute (ODI), London, UK.

Chapter 7
Rights-Based Approach to Livelihoods

Abstract The rights-based approach (RBA) to development offers a new perspective to traditional development by bringing in the role of law and human rights. It challenged the welfare state and market-led models of developmental thinking, by looking at poverty not as a fault of the individual or the absence of goods and services, but as the result of social injustice. It is normatively based on the three human rights principles of participation, accountability, and non-discrimination. The central tenet of RBA is about identifying the root causes of poverty and looking at basic needs as human rights. It aims to empower right holders to claim their rights and enable duty bearers to fulfil their obligations. Hence applying the RBA in practice will involve a change in policies, laws, programmes, and budgetary allocations in favour of the disadvantaged groups and will entail mobilizing people, report right violations, and petitioning courts to intervene. Experiences with implementing RBA show that this approach can be valuable in bringing structural issues related to marginalization and discrimination to the fore. We opine that the RBA adds real value and additionality to the family of rural livelihood approaches and should be used strategically to challenge deep-rooted injustice.

Keywords Welfare state · Market-led approach · Injustice · Discrimination · Barriers · Opportunities · Entitlements · Structural barriers · Capabilities · Freedom · Charity · Duties

7.1 Evolution of the Rights-Based Approach to Development

The twentieth century was dominated by two schools of development thinking—the welfare state and the neoliberal market-led approach. The welfare state model defined poverty as the absence of specific public goods or services or technical know-how and hypothesized that provisioning the same by the government would result in poverty reduction and development (Offenheiser & Holcombe, 2003). It assumed that development aid and market forces would 'lift all boats' and address the systemic problems of poverty and marginalization (Deka, 2012). However, after experimenting with these approaches for more than five decades, while the overall poverty levels have come down, but studies show that the gap between the rich and the poor is widening, wealth is getting concentrated in the hands of a few, and certain sections of the society are still caught in poverty traps (Deka, 2012). A significant drawback in the model of a welfare state is that it lacks a mechanism to hold the governments and duty bearers accountable for their action, or lack thereof (Offenheiser & Holcombe, 2003). While some progress was made, the structural issues related to poverty were not addressed and the poor were looked upon as beneficiaries rather than as equal partners (de Silva, 2013). In this new approach to development, what was earlier thought of as 'basic needs', are in fact 'human rights', and denying them would lead to human rights violations (Rand & Watson, 2007). For example, lack of access to safe drinking water is a violation of economic and social rights, while the fear of speaking out about this problem is a violation of civil and political rights (Rand & Watson, 2007).

S. Tambe, *Teaching and Learning Rural Livelihoods*, Sustainable Development Goals Series,
https://doi.org/10.1007/978-3-030-90491-3_7

Table 7.1 How is the rights-based approach different? (Adapted from Kirkemann Boesen & Martin, 2007)

Parameter	Charity approach	Needs approach	Rights-based approach
Goal	Goal is on deprivation and increasing charity	Goal is on meeting the needs	Goal is on realizing rights
Focus	Focus on input, not outcomes	Focus on input, output, and outcomes	Focus on process and outcomes
Orientation	Service oriented	Development oriented	Human development
Rationale	Recognizes moral responsibility of rich towards poor	Recognizes needs as valid claims	Recognizes rights as legal claims to be fulfilled by duty bearers
Outlook	Individuals are seen as victims	Individuals are objects of development interventions	Individuals have rights and are legally entitled to assistance
Strategy	Focusses on the manifestation of problems	Focusses on immediate causes of problems	Focus on structural causes of problems
Participation	Informing	Consultation	Citizen control
Impact	Changes condition of people	Changes in development status of people	Changes position of people
Political support	Lack of political will is an explanation	Lack of political will is an explanation	Lack of political will is a failure of duty

This led to the growth of the rights-based approach (RBA) to development that explains poverty as an outcome of injustice and includes marginalization, discrimination, and exploitation as the central causes of poverty (Kirkemann Boesen & Martin, 2007). It led to the growing realization that poverty is more than an economic need, and the state needs to address the more complex and fundamental causes of poverty (Box 7.1). This approach strives to identify the key barriers that prevent people from accessing opportunities and enhancing their wellbeing (Offenheiser & Holcombe, 2003). The poor are not perceived as helpless and dependent on government aid for escaping poverty, but as right holders who have entitlements and advocate for a fair allocation of public goods and services (Deka, 2012). The focus then shifts away from the simple needs-based approach to lowering these structural barriers that impede communities from exercising rights, building capabilities, enjoying freedoms, and receiving fair treatment. How the rights-based thinking differs from the traditional charity and needs-based approaches is illustrated in Table 7.1).

Box 7.1 The Complexity of Poverty

Our understanding of a situation guides our response to address it. Let us take the case of a poor rural household facing chronic food insecurity. Their poverty can be looked upon as a lack of adequate food and the solution can be provided them with food. If the cause of this food insecurity is attributed to the small plot of marginal land they cultivate, then more solutions come to mind such as agriculture intensification, livelihood diversification, increasing the land holding, and others. If further probing reveals that the cause of their penury is the alienation of a large chunk of the household's land into the hands of the village money lender when the household could not repay the usurious loan on time, then the solution is different again. If we delve deeper and find that the money lender belongs to a higher caste, is influential in local politics, and the household belongs to a community that has been historically discriminated against, then the complexity deepens further. Hence, persistent probing and asking 'why' is the key. We note that simple solutions only provide symptomatic solutions to the complexity of poverty, which often has structural causes.

Source: Adapted from Kirkemann Boesen and Martin (2007)

This right to development discourse gained currency with the end of the Cold War, NGOs campaigning for a broader array and importance of human rights, and the need for deeper participation and democratic decision making (de Silva, 2013). This realization that civil and political rights are necessary but not sufficient to address poverty resulted in the RBA to development being embraced by governments and developmental agencies (de Silva, 2013). Before this, development was looked upon as a project within international cooperation and considered as the terrain of economists, while human rights were seen as a separate domain occupied by lawyers and activists (Cornwall & Nyamu-Musembi, 2004). Sen (1999) in his seminal work on 'Development as Freedom' argued that freedom is both the primary objective of development and the principal means of development, thereby making their inseparability clear (Box 7.2). He argued that poverty should be understood as a multidimensional condition, shaped by constraints on individuals' freedoms and capabilities, and perpetuated by discriminatory institutions and failed political systems (Nelson, 2007) (Box 7.3).

Box 7.2 Development as Freedom

The twenty-first century heralded a new paradigm in development thinking with development and rights becoming intertwined, as different strands of the same fabric. Development came to be re-defined in terms of human rights as a constitutive part, leading to the reframing of the goal of ending poverty as a legal obligation. The champion of this new paradigm was Amartya Sen who defined development as the expansion of capabilities or substantive human freedoms, and the capacity to lead the kind of life that one has reason to value. He advocated the lowering of barriers that obstructed freedom such as lack of economic opportunities, political liberties, social acceptance, neglect of public facilities as well as intolerant or overbearing states (Sen, 1999). According to Sen, 'Freedom is both the primary objective of development and the principal means of development. Development is the process of expanding human freedom that allows people to lead lives that they have reason to live'. These ideas have since then made great inroads in the international development discourse.

Source: Uvin (2007)

Box 7.3 Amartya Sen: Does Democracy Avert Famine?

Amartya Sen was awarded the Nobel Prize in 1998 in Economic Sciences for his work on welfare economics and social choice theory. He was born in Santiniketan, Bengal, in India. His father was a professor of chemistry in Dhaka (now part of Bangladesh), where he also received his first education. He was further educated at Presidency College in Calcutta, India (now Kolkata) and later went on to study at Cambridge, UK, where he received his Ph.D. in 1959. He taught economics at multiple institutions across India and at Oxford and Cambridge universities, as well as in the USA, including at Harvard University. In his seminal work, he challenged the prevailing notion that famines are a result of food supply shortages, and demonstrated that they are instead caused by entitlement failure. His interest in famine stemmed from a lived experience when as a 9-year-old boy, he saw the Bengal famine of 1943 closely, in which nearly three million people perished. He concluded that this staggering loss of life was unnecessary, as there was adequate food supply at the time, but its distribution was limited as rural labourers who had lost their jobs lacked the ability to purchase the food. In his book 'Poverty and Famines: An Essay on Entitlement and Deprivation' (1981), Sen revealed that in many cases of famine, food supplies were not vastly reduced. Rather, several social and economic factors like declining wages, rising food prices, unemployment, and poor food distribution systems had led to starvation among certain groups in the society. His studies on famines helped create a deeper

(continued)

Box 7.3 (continued)

understanding of the economic reasons behind famine and poverty. He also argued that social reforms such as education and public health must pave the way for economic reform. Sen's views have guided policymakers to explore ways to replace the lost income of the poor, with public works projects, for example, and to maintain stable food prices to keep it affordable to the vulnerable groups. Governments and international organizations around the world continue to be vastly influenced by his ground-breaking work on famines, welfare economics, and beyond.

Source: Britannica, T. Editors of Encyclopaedia (2021)

This growing convergence between development and human rights enabled the RBA to become inclusive of poverty (de Silva, 2013). The concerns developed from a first generation negative concern with protecting individual civil and political (CP) rights, to a broader and more developmental concern with ensuring economic, social, and cultural (ESC) rights linked to poverty reduction goals (Blackburn et al., 2005). Hence, this approach sets the achievement of human rights as the objective of development, and thus requires the law to play an important role, and the need to promote legal rights and legal capacity building within the context of development (Broberg & Sano, 2018).

Consequently, in 1986, the UN General Assembly adopted the Declaration on the Right to development, which was reaffirmed by the World Conference on Human Rights in Vienna in 1993, and then at the 1995 World Social Development Summit at Copenhagen (de Silva, 2013). This declaration recognizes development as a human centred, participatory process and links human development to the realization of international human rights obligations (Filmer-Wilson, 2005). It states that 'the right to development is an inalienable human right under which every human person and all people are entitled to participate in, contribute to, and enjoy economic, social, cultural, and political development, in which all human rights and fundamental freedoms can be fully realized'. Soon the United Nations (UN), International NGOs, and the bilateral funding agencies decided to mainstream this approach in all their developmental programming (Broberg & Sano, 2018). The UN agencies define the RBA to development as a conceptual framework for the process of human development that is normatively based on international human rights standards and operationally directed to promoting and protecting human rights (Broberg & Sano, 2018).

7.2 Key Principles of the Rights-Based Approach to Development

The rights-based approach is grounded in the International Bill of Rights, the UN Universal Declaration of Human Rights, and other internationally agreed upon instruments, which provide globally recognized standards for what it means to live in dignity (Rand & Watson, 2007). The key distinctive feature of the RBA to development is that it starts by setting out a vision of what *ought* to be and then providing a normative framework to orient development cooperation (Cornwall & Nyamu-Musembi, 2004). This normative framework for the realization of human rights in the development context stands on the six human rights principles of universality and inalienability, inter-dependence and inter-relatedness, indivisibility, accountability, and the rule of law, participation and inclusion, and equality and non-discrimination. Broberg and Sano (2018) identify six key characteristics that define the RBA. First is the "concept of rights" that development is not charity but an attempt to realize rights and support right holders in claiming their rights. The second characteristic is that it makes sense to talk of a 'right' only when there is a corresponding duty bearer obliged to fulfill this right. Thus, a well-functioning and responsive state is often a prerequisite for the RBA to development to work. Third, when a donor applies the RBA to development, the focus is less on general service delivery,

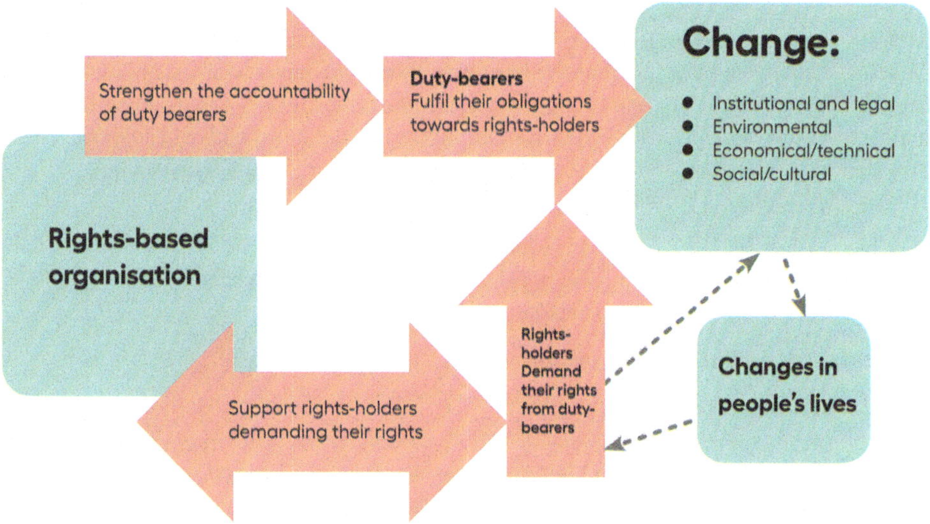

Fig. 7.1 How is the rights-based approach different? (Source: Kirkemann Boesen & Martin, 2007)

instead, the focus is on enabling the right-holder (target citizen group) to demand their rights and enabling the duty bearer (the state) to respond to these claims (Fig. 7.1). The fourth characteristic of this approach is that it relies on the premise that discrimination and inequality are among the key causes of poverty. Fifth it recognizes that to translate 'participation' and 'inclusion' into reality, it is important to strengthen the capacity for autonomous action. Hence, activism and advocacy become important elements of this approach. The sixth important characteristic is that all the facets of development cannot be directly cast as secured rights. For example, while the right to information can help fight corruption, it cannot impact the power asymmetry amongst stakeholders and guarantee a corruption-free future.

7.3 Rights-Based Approach to Development in Practice

What livelihood approaches have lacked all along is identifying the root causes of the generation and perpetuation of poverty and vulnerability (Conway et al., 2002). The RBA can add value by providing insights into power relations which are the key determinants of poverty. Power takes different forms, is acted out in different spaces, and occurs at different levels (Box 7.4). The role of politics and power in the livelihoods of the poor is analysed using the policy, institutions, and processes (PIP) element of the DFID sustainable livelihoods approach (SLA). A power analysis can help in unpacking and examining the inner workings of the PIP element in relation to rural livelihoods. The RBA can reframe livelihood security as an entitlement and secure it as a law thereby putting the onus on the state to fulfil its obligation. It can build the capacity of right holders and duty bearers and thus re-politicize development by basing it on rights rather than on charity. The essential elements of RBA projects are listed in (Box 7.5).

Box 7.4 Addressing Power Relations to Achieve Sustainable Livelihoods

Achieving sustainable rural livelihoods and ending poverty requires addressing power relations. Unequal power relations are one of the main underlying causes of poverty and suffering. They manifest themselves in different ways, from unfair trade regulations between countries to social norms that discriminate against girls. Power can be conceptualized in many ways:

Power takes different forms:

- Visible: observable decision-making mechanisms
- Hidden: shaping or influencing the political agenda behind the scenes
- Invisible: norms and beliefs, socialization, ideology

Power is acted out in different spaces:

- Closed: decisions made by closed groups
- Invited: people asked to participate but within set boundaries
- Captured: less powerful actors claim a space where they can set their own agenda
- Deliberative: equal power sharing amongst all the stakeholders

Power occurs at different levels:

- Household, local, national, and global

We usually think of power as control over others, but power can also be subtle. Power can be:

- Power over: to keep someone weak, vulnerable;
- Power to: capability to decide actions and carry them out: knowledge, skills, tools.
- Power with: collective power through organization, solidarity, and joint action.
- Power within: personal self-confidence

Power analysis can help understand the underlying structural factors impeding efforts to reduce poverty, as well as the incentives for achieving sustainable livelihoods.

An in-depth analysis of the root cause of poverty reveals *why* rights are not being realized, *who* is responsible for fulfilling them, *what* aspects of the policy framework need to be amended to enable them, and *how* poor people can themselves play a key role in accessing their rights.

Source: Oxfam (2014), Rand and Watson (2007)

Box 7.5 "Essential" Elements of RBA Projects

1. Thorough analysis of underlying causes of poverty, including explicit and ongoing analyses of power, gender, and risk
2. Community-cantered development, including building sustainable capacity to claim rights and to drive decision making
3. Duty bearers engaged, strengthened, and held accountable
4. Advocacy for sustainable change in policy and practice
5. Alliance-building
6. Working at multiple levels (e.g. local, national, international)
7. Focus on groups that are marginalized and discriminated against
8. Problems framed as rights issues and linked to international, national, or customary standards

Source: Rand and Watson (2007)

Different actors play varied roles in the application of the RBA in practice. While NGOs may place considerable emphasis on advocacy, government-backed funding agencies may not employ this role, while UN agencies will do so in subtle ways such as indirectly supporting NGOs (Broberg & Sano, 2018). The translation of the RBA in practice begins with an analysis of the present status of human development, and the inequities in diverse groups of stakeholders being able to access their developmental rights (Broberg & Sano, 2018). This analysis includes capacity gaps that prevent right holders from exercising their rights, as well as how duty bearers can contribute and engage to redress violations. Broberg and Sano (2018) list three key components of a rights-based analysis. First is the assessment of the actual status of human rights in the setting where development assistance is to be provided. This includes an assessment of right deprivations, fulfilment gaps, and power analysis. The second is to identify the vulnerable groups who are targeted in the proposed intervention. The third is to build on human rights principles such as participation, accountability, and non-discrimination.

The RBA to development involves empowering marginalized stakeholders, countering injustice in the form of oppression and exclusion, and changing power relations, most of these tasks lie outside the legal arena, falling squarely in the political realm (Uvin, 2007). The enforcement of RBA to development is centred around advocacy and dialogue over livelihood rights that right holders may like to claim from duty bearers. Hence, the entire setup or implementation structure of RBA is very different from conventional development assistance. We illustrate this difference with an example from the state of Uttar Pradesh in India of empowering women in their struggle against violence (Box 7.6). The establishment of the *Sakhi Kendra* enabled the marginalized women to come together and advocate when wrongs were committed against women and girls (Broberg & Sano, 2018). Translating the RBA to development is best reflected in India's endeavour during the first decade of this century to translate the development goals of transparency, employment, education, food security, forest tenure, and others into a legal right for its citizens (Box 7.7).

Box 7.6 Domestic Violence and Discrimination Against Women

Uttar Pradesh is a state in north India where a majority of women suffer physical abuse and domestic violence. To counter this the women in Kanpur town came together to form the *Sakhi Kendra*, a not-for-profit organization working for 40 years now, focussing on alleviating inequality and social injustice. The overall goal of this rights-based organization is to address the underlying causes of poverty. Its focus is explicitly on the empowerment of women and girls, especially from the poor marginalized or socially excluded communities, leading to improvement in their lives and livelihoods. It assists in re-housing, legal support, and medical aid for women who have suffered physical abuse, rape, or involuntary prostitution. It has successfully dealt with 38,600 cases of violence against women. Also, it has organized sensitization programmes for the police and other enforcement agencies and has started reviewing the policies and laws related to domestic violence and discrimination in the state.

Source: Broberg and Sano (2018), www.sakhikendra.org

Box 7.7 Do Rights-Based Legislation Help?
In the first decade of the twenty-first century, India enacted legislation to turn development goals such as transparency, employment, food security, primary education, and responsive governance into legal rights for its citizens. It was envisaged that by providing a legal basis to development, the citizens would be able to hold the state accountable and demand effective delivery. But can mere legislation effect such a significant transformation? In the case of the Right to Education Act, despite regular court intervention, the states were not able to provision the infrastructure legislated in the Act. This failure resulted from limited administrative capacity, irregular funding support, staff positions lying vacant, administrative logjams, and working in silos. Under the Right to Employment Act, one radical entitlement provided to citizens was the right to conduct regular social audits of works. In social audits, citizens audit government books to verify whether what is recorded represents ground realities. Though several functionaries were found guilty of misusing funds, the complex disciplinary norms prevented prompt punitive action. Based on this experience, do we reject the right-based approaches to development as pure rhetoric? While there is no doubt that the vision of the rights-based approach is powerful and futuristic, achieving it will need investments in building administrative capability, improving staff capacity, better funding and training, efficient management systems, and streamlining disciplinary action.

Source: Aiyar (2013)

As we can note from these examples, the activities in a RBA to development will involve lobbying with decision makers to change policies, laws, programmes, and budgetary allocations to favour the disadvantaged groups. Activities that were once thought of as peripheral like research, advocacy, evaluation, public education, and organizational development will now occupy the centre stage. Because resources are limited, some funding will likely be shifted from direct services to the poor toward efforts that target the underlying causes of poverty (Offenheiser & Holcombe, 2003). It will also entail mobilizing people and building their capacities to demand these changes and use media to raise awareness, report abuses of power, and rights violations. Further, monitoring the delivery of government services, reporting human rights violations, functioning as a watchdog, creating ombudsmen, protecting whistleblowers, strengthening complaint mechanisms, and finally petitioning courts to intervene to achieve justice and equality will also need to be taken up (Uvin, 2007).

7.4 Challenges to Development as a Human Right

Embedding the RBA to development implies a paradigm shift in how the problem is perceived and tackled. It clothes the developmental intervention in a legislative format and breaks it into legal claims, duties, and obligations, and mechanisms for adjudicating the violation of the same (Uvin, 2007). This necessitates a shift from need-based to rights-based approach and looking at development not as charity but as an obligation, with implications on state policy and accountability (Uvin, 2007). The challenge for the RBA to development lies in being able to move beyond theoretical and rhetoric support for the integration of development and human rights, to programmatic and policy measures and challenge and stretch the mandate and structures of existing organizations (Nelson & Dorsey, 2003). The on-ground enforcement of the rights-based approach is challenging as it involves unravelling complex power equations, addressing oppression and discrimination, and organizing the deprived for social action (Deka, 2012).

The human rights perspective lays higher emphasis on civil and political rights and gives less priority to economic, cultural, and social rights (Deka, 2012). Civil and political rights are based on negative freedoms, such as the right to life, liberty, and speech, which the state must merely guarantee. In contrast, social, economic, and cultural rights are associated with positive freedoms that are not seen as natural rights, as they have budgetary implications (Sengupta, 2000). Hence, even in countries of the global north, human rights perspectives often lay less emphasis on these development rights as they are not naturally enforceable by law (Frediani, 2010). The centrality of the RBA is that they are based on the working of legal systems, and casting all aspects of development into justiciable rights through legislation is not easy (Deka, 2012). The mere designation of rights is not sufficient, to ensure their realization. For marginalized groups that lack group solidarity or forms of collective representation or networking with organizations that can lobby on their behalf, formal rights have little effect on livelihoods (Conway et al., 2002). Strengthening the capacity of the duty bearers (generally the state) is equally important, as is highlighted in India's experience with framing its development in RBA (Box 7.7). Bringing in meaningful change using the RBA to development would be difficult when the state is weak or when the judicial machinery and institutional mechanisms are too under-resourced to fulfil their obligations.

7.5 Lessons Learnt and Way Forward

The most far-reaching impact of integrating the RBA to development is that the economic, social, and cultural (ESC) rights if taken seriously would suggest a new framework for development (Nelson & Dorsey, 2003). It brings back politics in poverty discourses and refocusses the attention on the institutions and processes that determine development outcomes (Blackburn et al., 2005). A particular strength of the RBA contribution is in making human rights more operational in the terrain of development and the economy (Gready, 2008). Moreover, the RBA is highly relevant for the most marginalized sections of society as it brings on the table new strategies for empowerment such as the use of advocacy, social organization, and legal remedies (Broberg & Sano, 2018). The approach also has the potential to reach out to the other groups that are not vulnerable in a classic economic sense but perhaps subjected to physical or social vulnerabilities, like the sexual minorities that are often exposed to assault (Broberg & Sano, 2018). In this approach, participation takes the form of citizen control to challenge and change the institutions that govern their lives (Blackburn et al., 2005). A RBA aspires to bring about a more equitable sharing of power among the stakeholders (Cornwall & Nyamu-Musembi, 2004). That said, it remains to be seen what human rights or RBAs can deliver in practice. Implementation will depend on contextual factors as well as stakeholders who are right holders, duty bearers, and support agencies (Broberg & Sano, 2018). At a practical level, it is clear that the provision of rights in law is not enough to ensure that they are respected by élites or enforced by the state. To realize citizens' full economic and social rights as laid out in international law would require levels of public spending and institutional capacity which are not feasible in poor states (Conway et al., 2002).

The RBA to development involves policy and legal changes that provide governments with the option of claiming compliance to international standards without implementing meaningful reform, at least in the redistribution of power and resources which is at the heart of social justice (Grugel & Piper, 2009). However, the RBA does offer an opportunity to political parties to legislate their developmental commitments, thereby making it difficult for a new political regime to alter it subsequently. Development interventions framed as RBA often have a longer life, and when implemented well can help political parties create a developmental legacy that can yield mileage even when not in power.

The experiences with RBA in development point out that while this approach can be instrumentally useful, morally robust, and can bring issues related to discrimination and marginalization to the fore, but operationalizing it in practice to address injustice effectively is not easy in the face of entrenched structural inequalities (Grugel & Piper, 2009). The uptake of this approach may vary from one context to the other, and for rural areas, this approach may be less effective because of lower literacy levels, and the state being less present (Broberg & Sano, 2018). At times, for the poor and marginalized, alternate livelihoods may be an important entry point that reduces dependence and enables people to contest their unrealized rights more effectively. Hence, the RBA should not be viewed as a silver bullet to solve all developmental and social justice issues, rather its application must be strategic and targeted based on the sector, the stakeholders, and the duty bearers. It is not suitable for mainstreaming in all development and must be viewed as a part of a family of rural livelihood approaches and used strategically to challenge injustice, deprivation, and marginalization.

Additional Resources

1. Introduction to human rights-based approach, six lessons in digital learning mode, The Danish Institute for Human Rights https://www.humanrights.dk/learning-hub/introduction-human-rights-based-approach
2. Human Rights-Based Approach to Development in practice—a Kenyan activist's experience https://www.youtube.com/watch?v=5P93m2FlpTY
3. Integrating human rights, leave no one behind, and gender equality into UN Cooperation Frameworks https://www.youtube.com/watch?v=mHHy1gDn4x8

Exercises

1. The Scheduled Tribes and Other Traditional Forest Dwellers (Recognition of Forest Rights) Act, 2006 was enacted by the Indian parliament to undo the historical injustice meted out to the forest dwellers during colonial times when the community forests were nationalized labelling them as encroachers. Critically analyse the model of forest governance envisioned in this Act using the lens of rights-based approach. https://tribal.nic.in/FRA/data/FRARulesBook.pdf
2. The Mahatma Gandhi National Rural Employment Guarantee Act, 2005 (MGNREGA) was enacted by India to replace the Sampoorna Grameen Rojgar Yojana (SGRY) programme with a rights-based approach. This statute is hailed as the largest and most ambitious social security and public works programme in the world. Critically analyse the extent to which the principles of RBAs, namely participation, accountability, non-discrimination, transparency, and empowerment are inbuilt into this legislation. https://www.indiacode.nic.in/bitstream/123456789/15229/1/nrega_act_2005.pdf
3. Allison et al. (2012) discuss the role that rights-based approaches can play in enhancing the chances of achieving both human development and resource sustainability outcomes in small-scale fisheries of developing countries. Discuss the strategy they propose to graduate from fishing rights to human rights. https://www.researchgate.net/profile/John-Kurien/publication/229916505_Rights-based_Fisheries_Governance_From_fishing_rights_to_human_rights/links/5ab0fa79aca2721710febc17/Rights-based-Fisheries-Governance-From-fishing-rights-to-human-rights.pdf
4. What are the essential elements that make a project 'right-based'? How is the rights-based approach different from the needs-based approach? Discuss by building on the case studies discussed in Rand and Watson (2007) in the Appendix. https://oxfamilibrary.openrepository.com/bitstream/handle/10546/134967/bk-rights-based-approaches-300308-en.pdf?sequence=1&isAllowed=y
5. What are the contexts where the rights-based approach to development can add significant value in achieving the global goal of ending poverty and hunger by 2030? Discuss by building on the case studies discussed in Rand and Watson (2007) in the Appendix. https://oxfamilibrary.openrepository.com/bitstream/handle/10546/134967/bk-rights-based-approaches-300308-en.pdf?sequence=1&isAllowed=y

6. What are the contextual factors that can accelerate or impede the gains envisaged under a rights-based approach to development? Discuss by building on the case studies discussed in Rand and Watson (2007) in the Appendix. https://oxfamilibrary.openrepository.com/bitstream/handle/10546/134967/bk-rights-based-approaches-300308-en.pdf?sequence=1&isAllowed=y
7. As a consultant eager to embed the rights-based approach in the developmental strategy of a low-income country, what are the aspects you would like to include in your presentation to policymakers to highlight the value this approach can add?
8. Cameroon was one of the early countries in Africa to embark on a tenurial reform in forest governance. Study the journal paper of Duguma et al. (2018) and discuss the role of enabling factors in improving the performance of this rights-based approach to forest governance? https://www.jstor.org/stable/pdf/26799126.pdf
9. Oxfam launched the 'Make Trade Fair Coffee Campaign' in Ethiopia to change the policies and barriers that small holder coffee farmers faced to improve their livelihoods (Rand & Watson, 2007, Appendix A1, pp. 43–53). Discuss what elements of the rights-based approach were used in this campaign, and what are the emerging impacts. https://oxfamilibrary.openrepository.com/bitstream/handle/10546/134967/bk-rights-based-approaches-300308-en.pdf?sequence=1&isAllowed=y
10. Sanyal et al. (2015) describes how the Rural Livelihood Mission programme (JEEViKA) resulted in empowering poor women in the eastern state of Bihar in India. This region is characterized by high levels of poverty, patriarchy, and illiteracy coupled with gender and caste hierarchies that are oppressive. Describe the processes and mechanisms this project adopted to create the social impact despite the deep-rooted structural barriers it was confronted with.
11. CARE implemented the 'Locally Intensified Farming Enterprises and New Options for Pest Management' project in Bangladesh to assist food insecure households to build their capacities and adapt to new farm technologies (Rand & Watson, 2007, Appendix A3, pp. 63–71). Using this case study, explain how the rights-based approach differs from the needs-based approach? https://oxfamilibrary.openrepository.com/bitstream/handle/10546/134967/bk-rights-based-approaches-300308-en.pdf?sequence=1&isAllowed=y
12. CARE implemented the 'Economic, Political and Social Empowerment for Small Producers' in Guatemala's Central Highlands (Rand & Watson, 2007, Appendix A6, pp. 91–99). Using this case study, explain how the rights-based approach is helping change the power relations in society? https://oxfamilibrary.openrepository.com/bitstream/handle/10546/134967/bk-rights-based-approaches-300308-en.pdf?sequence=1&isAllowed=y
13. Make a comparative assessment of these three case studies of Ethiopia, Bangladesh, and Guatemala (Rand & Watson, 2007, Appendix A1, A3, A6) using the three parameters of 'use of right-based elements', 'impact made', and 'based on normative principles'. https://oxfamilibrary.openrepository.com/bitstream/handle/10546/134967/bk-rights-based-approaches-300308-en.pdf?sequence=1&isAllowed=y

References

Aiyar. (2013, April 11). Do rights-based legislations help? *Financial express*. Retrieved June 12, 2021 at 21:00, from https://www.financialexpress.com/archive/column-do-rights-based-legislations-help/1100707/.

Allison, E. H., Ratner, B. D., Åsgård, B., Willmann, R., Pomeroy, R., & Kurien, J. (2012). Rights-based fisheries governance: From fishing rights to human rights. *Fish and Fisheries, 13*(1), 14–29.

Blackburn, J., Brocklesby, M. A., Crawford, S., & Holland, J. (2005). Operationalising the rights agenda: Participatory rights assessment in Peru and Malawi.

Britannica, T. Editors of Encyclopaedia. (2021). Amartya Sen. *Encyclopedia Britannica*. Retrieved from https://www.britannica.com/biography/Amartya-Sen.

Broberg, M., & Sano, H. O. (2018). Strengths and weaknesses in a human rights-based approach to international development–an analysis of a rights-based approach to development assistance based on practical experiences. *The International Journal of Human Rights, 22*(5), 664–680.

Conway, T., Moser, C., Norton, A., & Farrington, J. (2002). *Rights and livelihoods approaches: Exploring policy dimensions*. Natural Resource Perspectives, The Overseas Development Institute.

Cornwall, A., & Nyamu-Musembi, C. (2004). Putting the 'rights-based approach'to development into perspective. *Third World Quarterly, 25*(8), 1415–1437.

de Silva, T. E. (2013). *Principles in international development: Sustainable livelihoods and human rights-based approaches*. Division of Economic and Labor Research Office of Trade and Labor Affairs U.S. Department of Labor, USA.

Deka, A. (2012). Local and global poverty: Insights using a rights-based approach. *Advances in Social Work, 13*(2), 471–483.

Duguma, L. A., Minang, P. A., Foundjem-Tita, D., Makui, P., & Piabuo, S. M. (2018). Prioritizing enablers for effective community forestry in Cameroon. *Ecology and Society, 23*(3).

Filmer-Wilson, E. (2005). The human rights-based approach to development: the right to water. *Netherlands quarterly of human rights, 23*(2), 213–241.

Frediani, A. A. (2010). Sen's capability approach as a framework to the practice of development. *Development in Practice, 20*(2), 173–187.

Gready, P. (2008). Rights-based approaches to development: What is the value-added? *Development in Practice, 18*(6), 735–747.

Grugel, J., & Piper, N. (2009). Do rights promote development? *Global Social Policy, 9*(1), 79–98.

Kirkemann Boesen, J., & Martin, T. (2007). *Applying a rights-based approach—An inspirational guide for Civil Society, Danish Institute for Human Rights*. Retrieved June 19, 2021 at 10:00, from https://dspace.ceid.org.tr/xmlui/bitstream/handle/1/1556/applying%20a%20rights%20based%20approach.pdf?sequence=1&isAllowed=y.

Nelson, P. J. (2007). Human rights, the millennium development goals, and the future of development cooperation. *World Development, 35*(12), 2041–2055.

Nelson, P. J., & Dorsey, E. (2003). At the nexus of human rights and development: New methods and strategies of global NGOs. *World Development, 31*(12), 2013–2026.

Offenheiser, R. C., & Holcombe, S. H. (2003). Challenges and opportunities in implementing a rights-based approach to development: An Oxfam America perspective. *Nonprofit and Voluntary Sector Quarterly, 32*(2), 268–301.

Oxfam. (2014). *Quick guide to power analysis, oxfam policy & practice*. Retrieved June 13, 2021 at 10:00, from https://policy-practice.oxfam.org/resources/quick-guide-to-power-analysis-313950/.

Rand, J., & Watson, G. (2007). *Rights-based approaches learning project*. Oxfam America and Cooperative for Assistance and Relief Everywhere, Inc. (CARE), distributed by Oxfam GB, Boston. Retrieved June 22, 2021 at 11:25. from https://oxfamilibrary.openrepository.com/bitstream/handle/10546/134967/bk-rights-based-approaches-300308-en.pdf?sequence=1&isAllowed=y.

Sanyal, P., Rao, V., & Majumdar, S. (2015). *Recasting culture to undo gender: A sociological analysis of Jeevika in Rural Bihar, India*. World Bank Policy Research Working Paper, (7411). Retrieved June 16, 2021 at 11:00, from http://crossasia-repository.ub.uni-heidelberg.de/3789/1/India%20Recasting%20Culture.pdf.

Sen, A. K. (1999). *Development as freedom*. Alfred A. Knopf.

Sengupta, A. (2000). Realizing the right to development. *Development and Change, 31*(3), 553–578.

Uvin, P. (2007). From the right to development to the rights-based approach: How 'human rights' entered development. *Development in Practice, 17*(4–5), 597–606.

Chapter 8
Graduation Approach for the Ultra-poor

Abstract The need for the graduation approach is built on two key realizations—first that the ultra-poor were not being benefitted by the mainstream poverty-reduction programmes, and second that the poor are not homogenous and there is a need to look at the ultra-poor as a separate category with specific needs different from the poor. The design of poverty-reduction programmes faced polarized debates on the neoliberal trickle-down approach versus the interventionist approach, social protection versus livelihood promotion, asset transfer versus microfinance, and the like. The BRAC graduation approach integrates these dualities by combining a sequenced, step-by-step, time-bound set of interventions, customized to the needs of the ultra-poor, starting from protection to financial inclusion to promotion, and finally engagement with markets. This approach was tested in multi-country trials and showed encouraging results and is being scaled up. Contextual factors such as cooperation from men, weak economic activity, low market demand, access to affordable health facilities, and others are crucial. The main challenges in scaling up are affordability, errors of inclusion, and sustaining the livelihood gains. We opine that this approach can play a significant role in ending extreme poverty and meeting the SDG goals of ending poverty and hunger.

Keywords Ultra-poor · Microfinance · Basic needs · Consumption smoothening · Financial inclusion · Skill development · Income generation · Asset transfer · Life coach · Mentoring · Convergence

8.1 Ultra-poor Getting Left Behind

Matin and Hulme (2003) point out that in recent years, there have been three significant advances in the ideas that have impacted poverty reduction thinking. First, is the recognition that the poor are not a homogenous group, such as landless workers, small farmers, women-headed households, etc., but have many different characteristics and thus will require a more nuanced understanding while refraining from broad-brush generalist approaches. Second, that the debate between the efficacy of protection approaches versus promotion approaches in poverty reduction is increasingly recognized as sterile. It is now clear that sustainable poverty reduction requires both a protection component (food aid, cash transfer, social assistance, etc.) and a promotional component (income generation, skill building, enhancing productivity, etc.). Third, is the understanding that the agency of the poor people must be seen as the central goal of poverty reduction and programmes that are prescriptive or top-down are likely to fail.

The ultra-poor residing in rural areas is the poorest section of the population who are assetless, vulnerable to shocks, and has wage labour as their only source of income (Halder & Mosley, 2004). These households do not own any agricultural land or even homestead, have no savings, face food insecurity, and sell manual labour with no other means of income. Hashemi and De Montesquiou (2011) define the ultra-poor as those households who despite spending 80% of their total expenditure on food, still cannot meet 80% of their standard calorie needs. These ultra-poor households are

S. Tambe, *Teaching and Learning Rural Livelihoods*, Sustainable Development Goals Series,
https://doi.org/10.1007/978-3-030-90491-3_8

chronically food insecure and most vulnerable to health shocks and natural calamities. They are largely unaffected by the economic growth and trickle down policies, remain socially marginalized, geographically isolated, many are female-headed households, have weak social capital and networking, remain beyond the reach of government schemes and services, and often lack hope for the future, and confidence in their abilities (Dharmadasa et al., 2015). Also, as Hashemi and De Montesquiou (2011) point out, development programmes including microfinance programmes have typically not been able to reach the poorest, especially those at the bottom of the developmental ladder. Hence, there was a felt need to disaggregate the poor and look at the ultra-poor as a unique category having special needs very different from that of the poor.

The traditional poverty headcount ratio is insensitive to the large variation in living standards among those living below the poverty line (Sen, 1976). Hashemi and De Montesquiou (2011) note that amongst those who live in extreme poverty as defined by the World Bank at USD 1.90/day, there are those at the very bottom who are unable to meet even their most basic needs. While pockets of extreme poverty exist around the globe, those deepest in poverty reside in sub-Saharan Africa (Fig. 8.1) (Cruz et al., 2015). There is thus a need to shift the policy discourse towards the plight of the ultra-poor and their special needs in this region that is trailing.

8.2 The Genesis of the Graduation Approach

Sir Fazle Hasan Abed founded the Bangladesh Rural Advancement Committee (BRAC)—a Bangladesh-based NGO in 1972 (Box 8.1). BRAC pioneered the graduation approach based on several decades of development work, especially when it realized that many of its interventions did not sustainably impact the ultra-poor who needed it the most (Box 8.2). Traditional poverty reduction

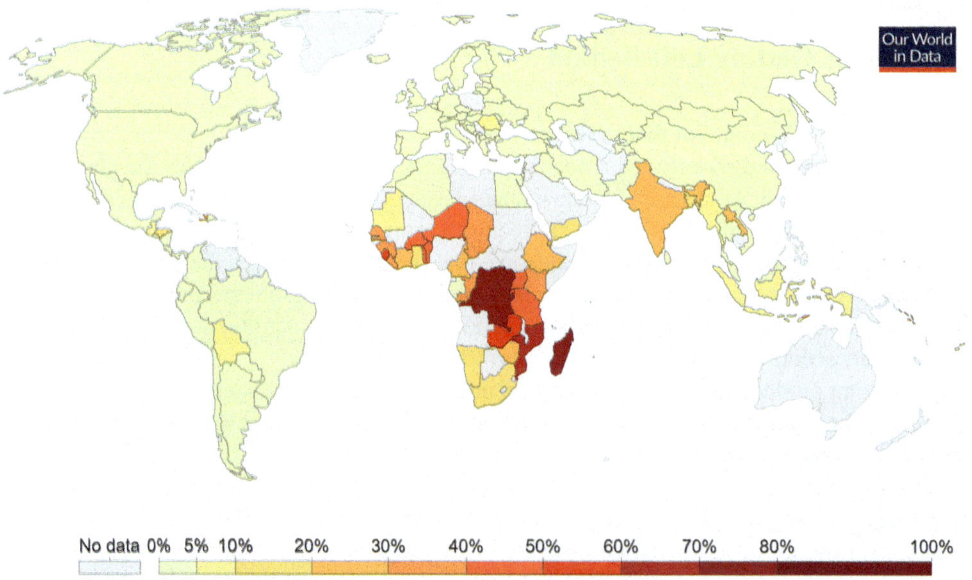

Fig. 8.1 Distribution of extreme poverty around the world. (Source Roser & Ortiz-Ospina, 2017)

programmes such as social protection, microfinance, financial literacy, and skill development inadvertently failed to reach the poorest and create lasting gains. The graduation approach is a comprehensive, intensive, time-bound, and sequenced set of interventions to lift the ultra-poor to sustainable livelihoods (Dharmadasa et al., 2015). It is comprehensive in the sense that it integrates the approaches of social security, livelihood development, and microfinance. It is intensive as it focusses on sizeable investments at the household level. It is time-bound as it commits to a time frame of 18–36 months for lifting the ultra-poor households out of poverty. It provisions tailored project components related to targeting, consumption support, savings, financial inclusion, skill transfer, asset transfer, life skills coaching, and microfinance in a sequenced manner.

Box 8.1 Sir Fazle Hasan Abed, Founder of BRAC

Sir Fazle Hasan Abed was born in 1936 in Baniachong, located in what is present-day Habiganj District of Bangladesh in a politically influential family. At the age of 18, he joined the University of Glasgow, Scotland to study naval architecture in a bid to do something radically different. Realizing that there was virtually no scope in ship building in East Pakistan (now Bangladesh), he joined the Chartered Institute of Management Accountants in London, thus completing his professional education in 1962. He acquired British citizenship but soon returned to East Pakistan (now Bangladesh) to join the oil company Shell, becoming Head of Finance in the local subsidiary. In 1970 he returned to Britain due to increasing violence as his country struggled for independence and campaigned to end the bloodshed. Arriving back in the newly independent Bangladesh at the end of 1971, he found the country devastated by war and the aftermath of cyclone Bhola of 1970. The cyclone remains the deadliest tropical cyclone ever recorded and one of the world's deadliest natural disasters with fatalities of over three hundred thousand people. There were few functioning schools, health facilities, agricultural resources, and the economy was in ruins.

This experience had a profound effect on him and he soon established BRAC (Bangladesh Rural Advancement Committee) to address the needs of the poor with a desire to develop their capacity and manage their lives in a better way. To evaluate the underlying causes of poverty, he realized the need for a multidimensional approach to address issues around sanitation, agriculture, education, hygiene, and family planning, among others. Today BRAC is one of the largest international development NGOs in the world, operating across a dozen countries in Africa and Asia. Their primary objective is to empower individuals living in poverty and inequality. In 2019, BRAC was ranked first among the world's top 500 NGOs by Geneva-based NGO Advisor; based on its impact, innovation, and sustainability, fourth year in a row. Sir Fazle has been honoured with numerous national and international awards for his achievements in leading BRAC and for his contributions to social development. He was an influential voice from the Global South in the International development circles. He passed away in 2019 leaving behind a legacy of strong leadership in the world of development.

The term 'graduation' is based on the analogy that just like students pass out from a university to seek future livelihoods after acquiring the necessary knowledge and skills, in the development discourse providing a holistic package to the extremely poor households would result in them leaving social protection after attaining a certain wellbeing status (Devereux & Sabates-Wheeler, 2015). Its theory of change is based on the recognition that a single intervention such as cash or asset transfer is unlikely to achieve a sustainable impact on the livelihoods of the ultra-poor, and that a holistic support package has a greater likelihood of constructing pathways out of poverty that leads towards self-reliance (Hashemi & Umaira, 2011).

Box 8.2 IGVGD Programme in Bangladesh
The BRAC (Bangladesh Rural Advancement Committee) was founded in Bangladesh in 1972 and has been involved in the successful implementation of large-scale and diverse rural development programmes across sectors such as microfinance, health, education, water, sanitation, and livelihoods. In 1985, BRAC in order to support the poorest started the Income Generation for Vulnerable Group Development (IGVGD) programme jointly with the World Food Programme. Other than direct food transfers the programme also provisioned skill development for income generation and access to financial services. While the participants showed progress in income generation, but most could not sustain the livelihood trajectory as they had low aspirations, dependence on food subsidy, and lacked confidence in applying the newly acquired skills. This led to the realization that often development programmes do not reach the poorest and impact them sustainably. Also, often the development interventions such as social protection measures, financial inclusion, and livelihood development are piece-meal and lack integration and sequencing. They realized that while the poor were able to embark on a sustainable development trajectory, the ultra-poor or the poorest were unable to use these interventions to escape poverty and hence continued to stagnate. This led to the realization that a separate set of interventions are needed to address the unique needs of the ultra-poor to lift them out of poverty.

Source: Dharmadasa et al. (2015)

8.3 Key Elements of the Graduation Approach

The graduation approach is based on the premise that poverty is multidimensional and will need a basket of poverty reduction interventions carefully tailored and sequenced to meet the unique needs of the ultra-poor to have a lasting impact (Box 8.3). The approach envisions consumption support in short term (through food or cash transfer) while developing sustainable livelihoods in the long run (Schelzig & Rawal, 2020). The graduation programme targets the poorest households in a village and provides a productive asset grant, training, consumption support, life skills coaching, access to savings accounts, and health services (Banerjee et al., 2015). It is based on the premise that a combination of these activities is necessary and sufficient to establish sustainable self-employment activities and generating lasting improvements in their wellbeing (Banerjee et al., 2015). It hinges on the importance of asset accumulation to human wellbeing and in achieving livelihood outcomes (Sabates-Wheeler et al., 2018). It stands on the four pillars of meeting basic needs, income generation, finance and savings, and social empowerment (Fig. 8.2).

The first step of the approach is targeting the ultra-poor households by excluding the others. This is done using a combination of methods by first conducting participatory wealth ranking (PWR) to develop the criteria to identify the ultra-poor households and then applying it in a participatory manner to select them. This is then followed by cross-verification by senior programme staff who visit all the selected households and triangulate information from different sources before finalizing the selection list (Hashemi & De Montesquiou, 2011). The poorest are typically those households who are assetless, vulnerable, and suffer from multiple deprivations with wage labour being their only source of income. There are two types of errors in beneficiary selection—Type I and Type II (Van Domelen, 2007). In the Type I error or errors of inclusion, non-poor individuals are included in the programme due to inaccurate eligibility specification, elite capture, etc., also known as 'leakage' of transfers to the non-poor. In the Type II errors or errors of exclusion the poor are excluded from programme benefits due to budgetary limitations, geographical delimitations of programme scope, lack of outreach to inform the poor of a programme, etc. resulting in a lower coverage of the poor. While errors of wrong-

Box 8.3 The Five Building Blocks of the Graduation Approach
The typical programme components are as follows:

- **Social transfers** of cash or food to enable the households to meet their immediate consumption requirements and tackle food insecurity
- **Financial inclusion and savings habit** to inculcate financial discipline
- **Technical and business skills training** suited to build capacity to take care of the asset and generate incomes
- **Asset transfer** selected by the households from a basket of options that are short-listed based on local market survey and value chain analysis
- **Regular coaching and mentoring** by a trained facilitator through regular home visits

Adapted from Schelzig and Rawal (2020)

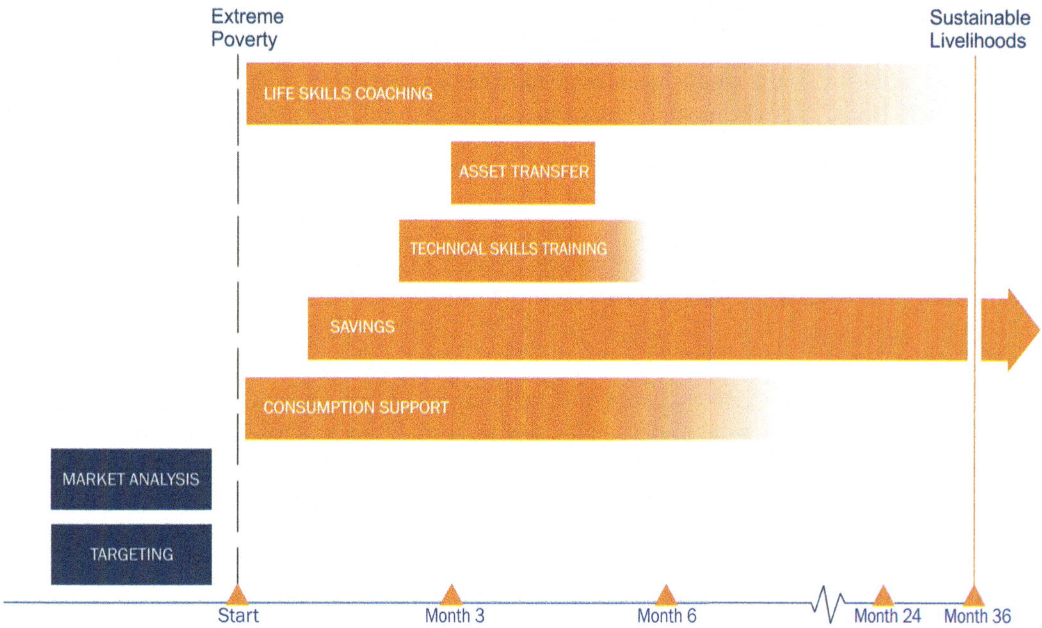

Fig. 8.2 Graduation to sustainable livelihoods. (Source: De Montesquiou et al., 2014)

ful inclusion (Type I) are easier to detect, exclusion of deserving households (Type II) is more challenging to tackle. Also, Type II errors are more serious as deserving households are left out.

The second step is the provisioning of consumption support as a cash stipend or food for a time-bound period. The main purpose of providing this safety net is to help families stabilize their food consumption until incomes from new income generating activities start coming in (El-Zoghbi et al., 2009). Households struggling with meeting their food security need typically develop tunnel vision, i.e. they are constantly worried about making ends meet in the short term and let mid-term or long-term opportunities slip by. Getting these households to work towards strengthening their long-term livelihood security is next to impossible unless their immediate needs are first met. In this context, consumption support is vital to relieve the households from the stress of day-to-day survival, so that they can focus on livelihood opportunities and plan for the future. In development practice, there are umpteenth examples of asset transfer initiatives gone awry as the poor households who received these assets found more value in liquidating them to meet their pressing needs rather than converting them into productive income generating assets in the mid-term.

The third step is savings, where the households are encouraged to save a part of their earnings and build financial discipline and become financially literate. The savings are expected to help them manage risks and tide over periods of inclement weather, ill-health, or lack of work opportunities. These savings also ensure that households do not have to sell their assets to meet their survival needs. Provisioning of consumption support and savings is a prerequisite before assets are transferred, as they ensure that the household will not sell the assets to meet the emergent, survival needs of the household in the face of shocks and stresses (El-Zoghbi et al., 2009).

The fourth step is capacity building of the households where they receive specialized skills on how to manage the transferred asset and operate a successful business (El-Zoghbi et al., 2009). This is followed by asset transfer wherein households are offered a menu of viable economic activities to choose from based on their experience and aspirations. This menu is developed based on careful market studies that assess space for new entrants into the value chain and scope for future expansion (Dharmadasa et al., 2015). This transfer of productive assets helps to kick-start their livelihood journey. The participants receive continuous handholding throughout the project period from a life coach who makes regular visits to the family guiding them, motivating them, counselling them, and helping them address the problems. These facilitators take on the role of a coach or mentor and develop deep bonds with the households. They also monitor the progress of the household to ensure that they are on track, guide in business plan preparation, enable social support, and encourage positive behavioural changes (Hashemi & De Montesquiou, 2011).

8.4 Graduation Approach in Practice

Banerjee et al. (2015) conducted a multi-country randomized trial with treatment and control groups of a graduation programme led by the Ford Foundation and Consultative Group to Assist the Poor (CGAP). This impact assessment found significant gains on key aspects, namely consumption, household assets, and food security, which was sustained 1 year later as well (Box 8.4). A 7-year evaluation of the programme in Bangladesh found a noticeable increase in annual earnings, productive working hours, consumption expenditures, and savings (Balboni et al., 2015). Sabates-Wheeler et al. (2018) found that different households respond differently and achieved varied outcomes to the same set of programme inputs as heterogeneity existed at multiple levels. They suggest that graduation programs will need to deliver different types of support for varied lengths of time to account for the diverse contexts if the agenda of 'leave-no-one-behind' has to be met.

Sabates-Wheeler and Devereux (2011) analysed the design of three well-known large-scale graduation programmes, namely the 'Challenging the Frontiers of Poverty Reduction: Targeting the Ultra-Poor (CFPR/TUP) in Bangladesh', 'The Productive Safety Net Programme (PSNP) in Ethiopia' and the 'Vision 2020 Umurenge Programme (VUP) in Rwanda'. They reported that the CFPR/TUP identified two key features that contributed to the success. Firstly, along with the assets the programme also provided a cash stipend for 18 months, and secondly, the assets were carefully selected and linked with related skill development. The allowance ensured a regular stream of income till the asset became productive and ensured that it was not sold off to meet the pressing needs. This programme resulted in households enjoying a more diversified and stable income and between 2002 and 2005 the participants in extreme poverty fell from 89% to 59%. In the PSNP in Ethiopia, two levels of graduation were provided. The first was to enable consumption smoothening and the second to ensure asset accumulation. The VUP in Rawanda targeted the extremely poor and provisioned public works for those who were able to work and unconditional cash transfers for others. The component of financial services was also introduced to offer low-interest loans for productive enterprises. Buoyed

Box 8.4 Core Graduation Criteria in Bangladesh

Graduation occurs when households achieve economic and social advancement measured by several criteria over the course of 24 months. Criteria vary given the social and geographic context of the programme. Listed below is the core graduation criteria in Bangladesh.

- At least 3 sources of income
- Asset value doubled since initial transfer
- Household consumes nutritional meals at least twice a day with protein at least once a week
- Participant engaged in household decision making (e.g. asset purchase)
- Improvement in home condition (e.g. corrugated roofs)
- Attends social or community events
- Access to sanitary latrine and clean drinking water
- Additional requirements where applicable:
  - School aged children attend school
  - No under-age marriages
  - Use of family planning

Source: Bangladesh BRAC Programme

by the success of these graduation programmes, many countries and NGOs have launched similar programmes that simultaneously protect and promote livelihoods (Sabates-Wheeler & Devereux, 2011).

It is crucial to note that the per participant cost for the graduation approach varies based on the country from US$ 330–650 in India to US$ 1900 in Haiti and on an average, costs around US$ 550 per participant (Hashemi & De Montesquiou, 2011). This cost includes consumption support, asset transfer costs, staff salaries, and head office overheads for the total programme duration (Hashemi & De Montesquiou, 2011). This costing depends on the amount spent on the asset (25–33%), the duration of the consumption support (typically 10–12 months), and the participants to staff ratio determined by the population density in the project area (Hashemi & De Montesquiou, 2011). While pilot programmes may appear expensive with high per participant investment, scaling up the programme can bring in economies of scale. Also, introducing a graduation programme does not mean provisioning all the components in a stand-alone manner. It can be converged with existing schemes that provision cash stipends, asset transfer, capacity building, and wage employment. These programmes can be combined by fine-tuning, sequencing, and better targeting, with coaching and mentoring as the glue to integrate these diverse programmes (Schelzig & Rawal, 2020).

Graduation programmes should not be implemented in a stand-alone manner but should build on existing social assistance, and promotional programmes (Sabates-Wheeler et al., 2018). Kidd and Bailey-Athias (2017) in their study highlighted the high levels of inclusion errors where several ineligible participants were included in these programmes in some countries. They also pointed out that the gains in household consumption often look deceptively high in percentage terms. However, in absolute terms, they may not necessarily be large enough to help the recipient households graduate out of poverty. The study also cautioned against assessing impact just 1 year after the support has ended. This is because the ultra-poor households in the face of shocks may sell off their assets as a coping mechanism and consequently revert to their earlier status. They advocated that only long-term assessment of the programme impacts can ascertain its sustainability. Their research stresses that the results of the graduation programmes have been overstated and showcased to create an impression that the impacts are far more than what they are in reality.

8.5 Graduation Trajectories

Sabates-Wheeler et al. (2018) conceptualize the analytical framework for post-graduation project livelihood trajectories. They define a household to have graduated when in the absence of project support it can ensure food security for all the 12 months and also withstand modest shocks. They regard this as sustainable graduation and contrast it with threshold graduation which is an administrative benchmark that signals the point after which the household is no longer eligible for the programme. They contest the stylized representation of the BRAC graduation approach (Fig. 8.2) for its assumed linearity, wherein the ultra-poor are expected to move smoothly up the graduation pathway like an escalator. While in practice, the livelihoods of the poor are unpredictable, erratic, and often subjected to setbacks and shocks (Sabates-Wheeler & Devereux, 2011).

Their research also pointed out that not all households will experience positive change due to the challenges, limitations, and context in which they operate. Every household or group of households has a clear baseline and target threshold level. They categorize the graduation in the post-programme period into four broad categories based on their performance during the project period (12 months) and during a substantial post-project period (24 months) (Fig. 8.3). The 'improvers' or those who have sustained the positive change, 'decliners' are those that showed improvement during the project period but have fallen back but not below the targeting threshold level, the 'crashing out' as those who since the end of the transfer have declined below the baseline threshold levels and the 'late improvers' as those who showed modest improvements during the project period but have since risen above the target threshold levels post-project. So while the programme inputs are similar, the diversity in the graduation trajectory of the programme participants can be attributed to heterogeneity amongst the households and their circumstances that either accelerate or impede the progress of the households.

8.6 Challenges and Contextual Factors

The main challenge in scaling up the graduation approach and its large-scale adoption is that it is relatively expensive. The investment per household in terms of the cost of the consumption support, asset transfer, and manpower costs for targeting, skilling, and coaching can be substantial (US$ 550 per participant). The coaching component is intensive in terms of human resources and frequent

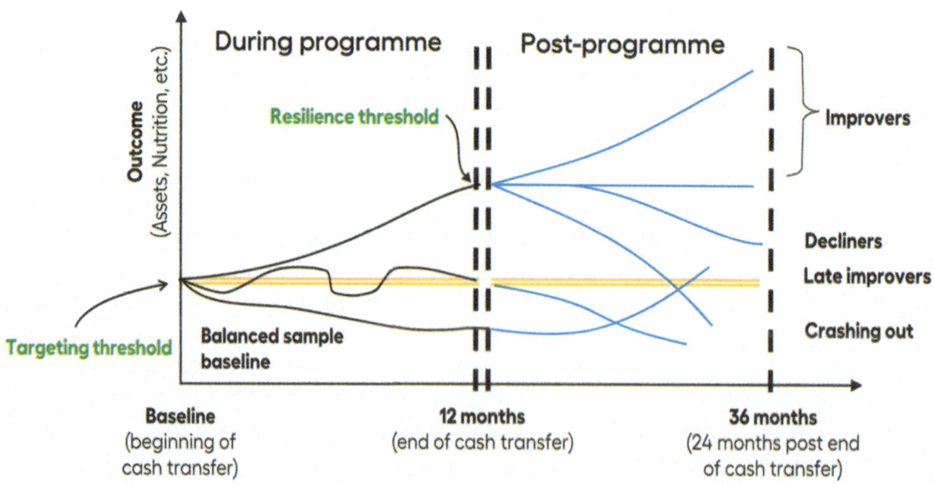

Fig. 8.3 Categorization of the graduation trajectories after cash transfer (CT) into improvers, decliners, later improvers, and crashing out. (Source: Sabates-Wheeler et al., 2018)

face-to-face sessions at the household level may need several more staff than what is normally provisioned in regular programmes (Schelzig & Rawal, 2020). Targeting only the ultra-poor and avoiding errors of inclusion of the non-ultra-poor can be challenging in specific settings. Also, generating political will and convincing policymakers to fund a programme that targets less than 8% of the population with high per capita costs can be challenging. However certain features on the other hand, such as ease in estimating the financial investment upfront (as it is based on the per household costing), being time-bound with a relatively short graduation horizon of 24–36 months, and having clear outcomes make it appealing to policymakers. The graduation programme should not be looked upon as a silver bullet to address the structural issues or root causes of poverty and vulnerability (Schelzig & Rawal, 2020). Another limitation is that it may not be suitable for ultra-poor households that are destitute, i.e. the elderly or the disabled who do not have the capacity for economic self-sufficiency (Schelzig & Rawal, 2020). The destitute are more suited for inclusion under a permanent social assistance programme rather than a graduation-oriented livelihood programme (Sabates-Wheeler et al., 2018).

Hashemi and De Montesquiou (2011) identified four contextual factors that are beyond the control of the project but impact its overall outcomes. As the programme operates at the household level, families where the men cooperated, and households that had access to social capital performed better than households that had a higher dependency ratio, those with abusive men or were women-headed. Sites that are economically depressed or remotely located provide fewer marketing opportunities to the participants to sell the products of their small businesses. Hence, an effort is needed to simultaneously expand the markets as well. Expenditure on health emergencies can be a major reason for households losing their savings or liquidating their assets. The existence of public health infrastructure providing affordable health services is crucial. Environmental challenges such as droughts, floods, natural calamities, and macro-economic, political, and economic crises also challenge the tenuous livelihood development of the poorest. Structural and systemic barriers such as discrimination, exclusion, oppression, and lack of accountability can also hamper the graduation of the households out of poverty.

8.7 Graduation and the Way Forward

The graduation approach is positioned as a strategy to target the ultra-poor who are unable to sufficiently leverage poverty-reduction programmes to escape poverty. This approach is being promoted as a key element in combating extreme poverty and in achieving SDG Goal 1 of ending poverty in all its forms everywhere. It is being scaled up in several countries of the global South and is being implemented not only in rural but in urban settings as well, targeting women as well as youth and is also being implemented in fragile and conflict-affected countries (Arévalo et al., 2018). It has also been adopted in diverse contexts in response to natural calamities, conflict situations, resettlement of vulnerable communities, etc. This approach can also be implemented in convergence mode, complementing the existing government programmes. Government adoption and implementation of graduation have picked up since 2015 (Box 8.5). Also, most of the government-led graduation approaches build on existing social protection programmes by adding productive livelihood features (Arévalo et al., 2018). It is expected that in the future this approach will continue to grow in scale and influence, with the support of donors and national governments (De Montesquiou & Hashemi, 2017). The key takeaways for the future are ensuring cost effectiveness, adapting to the local context, and ensuring sustainability of the livelihood trajectories.

Box 8.5 Sustainable Livelihoods Scheme of the Government of Bihar, India

Liquor was banned in the eastern state of Bihar, India in 2016. The ban was widely appreciated; however, it severely impacted the livelihoods of the households engaged in the value chain of country liquor. In August 2018, the Bihar government launched a new scheme called *Satat Jivikoparjan Yojana* (SJY) for the rehabilitation and empowerment of ultra-poor households engaged in production, transportation, and selling of country liquor through diversification of their livelihoods, capacity building, and improved access to finance. The scheme was also extended to the ultra-poor from communities who were discriminated against historically. SJY has a budget of about USD 120 million for the total programme duration of 3 years. The scheme aims to cover around one hundred thousand ultra-poor households with 24 months of livelihoods and mentoring support. The SJY programme is designed on the lines of the BRAC graduation approach and includes a holistic set of protection and promotion principles to provide a big push to the ultra-poor, over a short period of time, with the hope of unlocking the poverty trap. The scheme integrates a participatory identification process, asset-grant transfer, training, temporary consumption allowance, weekly mentoring, general life skills coaching, and access to savings account, and health information or services, all of which are integral components of the graduation approach. Building on the lessons and best practices of the graduation approach, the SJY scheme customized this approach in the context of Bihar.

Additional Resources

- For additional resources on the BRAC Approach and findings from pilots see the note titled, "Hashemi, S. M., & De Montesquiou, A. (2011). Reaching the poorest: Lessons from the graduation model. *Focus Note*, *69*(1), 1–15. available at the following website https://www.cgap.org/research/publication/reaching-poorest-lessons-graduation-model
- For comprehensive technical resources on the BRAC approach see the guide titled, "De Montesquiou, A., Sheldon, T., Degiovanni, F., & Hashemi, S. (2014). From extreme poverty to sustainable livelihoods: A technical guide to the graduation approach. CGAP and Ford Foundation." available at: https://www.cgap.org/sites/default/files/graduation_guide_final.pdf
- TED Talk Video of BRAC approach by Shameran Abed http://bracultrapoorgraduation.org/wp-content/uploads/2020/09/ShameranAbed_TEDTalk_subs.mp4
- Videos of BRAC field pilots in various continents
- **Africa**
  - The Graduation Approach: Ending Extreme Poverty in Rural Kenya https://www.youtube.com/watch?v=4RrbI0Np5Ag
  - The CGAP-Ford Foundation Graduation Program—REST Ethiopia https://www.youtube.com/watch?v=PnoOr7_Rnv8

South America

- Graduation programme in Ecuador https://www.youtube.com/watch?v=Mu1Ouo4gznM

Asia

- Livelihood support: opportunities for a better future in Bangladesh https://www.youtube.com/watch?v=dJNiTqf3uys

Exercises

1. What are the main lessons from the interview with Sir Fazle Hasan Abed that you would like to draw upon and embed in your end-poverty proposal? https://www.ultra-poverty.org/blog-post/graduating-from-ultra-poverty-an-interview-with-sir-fazle-hasan-abed/
2. For a developing country in the global South, study the basket of rural schemes and analyse to what extent they together provision all the BRAC elements of graduation and if there are any missing links. In these schemes are the destitute and ultra-poor categories being addressed separately and adequately?
3. 'Leaving no one behind' is the recurring and overarching objective of the 2030 global agenda. In response to this, has any country or donor agencies introduced new schemes or re-designed existing schemes to meet the two global goals, namely SDG 1—No Poverty and SDG 2—No Hunger. If so, discuss the design of these schemes.
4. Study the portfolio of livelihood development schemes of a developed European country, and analyse how they are different from a developing country in Asia or Africa.
5. Study the livelihood development schemes of one developing country each in Africa, Asia, and Latin America and analyse their similarities and differences?
6. Study the developmental programmes in a few Latin American countries and describe two growth promoting social safety nets that can prevent a household from falling deeper into poverty as well as improve the human development status?
7. Study the case study of the Kenya graduation pilot to assess the performance of the participants and the lessons learned. https://bracultrapoorgraduation.org/wp-content/uploads/2020/03/Kenya-Graduation-brief_updated-Feb-2020.pdf
8. Watch the video (https://www.youtube.com/watch?v=THXFhqsIdLQ) describing the BRAC graduation approach by Syed M. Hashemi. Based on this video and your understanding of this approach, what safeguards would you embed in a national poverty alleviation programme of a developing country that builds on the BRAC graduation approach?
9. Discuss whether the BRAC graduation approach is too expensive and will it lead to a dependency syndrome among the ultra-poor?
10. The BRAC Propel Toolkit in Chap. 6 discusses the learnings and innovations for scaling up. It advises convergence with national safety net programmes as a critical component for scaling up the graduation pilots. Analyse the considerations, roadblocks, and adaptions needed for paving a pathway to scaled up implementation. https://www.un.org/esa/socdev/egms/docs/2016/Poverty-SDGs/BRAC-PROPEL-Toolkit.pdf
11. Assess the rural development programmes of the global south to identify a few government schemes that recognize the ultra-poor as a separate category from the poor needing specific attention. Critically analyse how they take into account the high per unit cost, targeting challenge, the concern of elite capture, and the voicelessness of the ultra-poor.
12. Critically discuss whether governments of developing countries are generally reluctant to focus on the ultra-poor? Is it the high unit cost, targeting challenge, the concern of elite capture or marginality, and voicelessness of the ultra-poor that has resulted in this? Or is it that state governments are yet to recognize the ultra-poor as a separate category from the poor needing a specific set of inputs.
13. Most of the social safety schemes need domicile, voters card, residential proof, etc., and are provisioned by the federal or the state governments. How do nomadic tribes, inter-state migratory labour, beggars, and others access these schemes? Discuss the design improvements needed in these safety nets to enable the inclusion of these special category households.
14. Designing livelihood schemes for indigenous people, hunter-gathers, pre-agriculture tribal groups, and others is especially challenging as wealth accumulation is alien to their worldview. Discuss the design of livelihood development programmes that will cater to their needs.
15. Internal monitoring of the rural livelihood programme (JEEViKA) in Bihar, which is one of the poorest states in India, revealed that the ultra-poor households were getting left out. An in-house

learning booklet describes the steps taken to develop a separate strategy for the ultra-poor (World Bank Group, 2017, pp. 21–23). Discuss how did this ultra-poor strategy differs from the BRAC graduation approach? https://openknowledge.worldbank.org/handle/10986/29193?locale-attribute=en

16. Study how the BRAC asset-based approach was used to provision the creation of household livelihood assets under a public works programme (MGNREGA) in the journal paper of Tambe et al. (2019). Discuss what additional components were integrated into a public works programme to enable it to also create assets for the ultra-poor? What are the possible risks involved in this approach and how to address them? https://www.researchgate.net/profile/Sandeep-Tambe/publication/330483116_Coupling_a_ladder_to_the_safety_net_reinventing_MGNREGA_for_asset_creation/links/5c41fd50458515a4c72f8a0c/Coupling-a-ladder-to-the-safety-net-reinventing-MGNREGA-for-asset-creation.pdf
17. For eliminating extreme poverty, which of the following social sector strategies, namely livelihood development, unconditional cash transfer, or graduation perform the best. Study the journal paper by Sulaiman et al. (2016) to discuss the differences in design and impact of these social sector initiatives. https://www.poverty-action.org/sites/default/files/publications/Forum-Eliminating-Extreme-Poverty-Dec-2016.pdf

References

Arévalo, I., Kaffenberger, M., & de Montesquiou, A. (2018). *2018 State of Sector*. Synthesis Report, World Bank, Partnership for Economic Inclusion, Washington DC. Retrieved January 27, 2021 at 14:15, from https://www.findevgateway.org/paper/2018/11/2018-state-sector-synthesis-report.

Balboni, C., Bandiera, O., Burgess, R., & Kaul, U. (2015). Transforming the economic lives of the ultra-poor. In *IGC growth brief*. International Growth Centre.

Banerjee, A., Duflo, E., Goldberg, N., Karlan, D., Osei, R., Parienté, W., et al. (2015). A multifaceted program causes lasting progress for the very poor: Evidence from six countries. *Science, 348*(6236).

Cruz, M., Foster, J., Quillin, B., & Schellekens, P. (2015). *Ending extreme poverty and sharing prosperity: Progress and policies*. Policy Research Note PRN/15/03 (2015), World Bank Group, Washington, DC.

De Montesquiou, A., & Hashemi, S. M. (2017). The Graduation approach with social protection: Opportunities for going to scale. *Policy in Focus, 14*(2), 17–21.

De Montesquiou, A., Sheldon, T., Degiovanni, F., & Hashemi, S. (2014). *From extreme poverty to sustainable livelihoods: A technical guide to the graduation approach*. CGAP and Ford Foundation.

Devereux, S., & Sabates-Wheeler, R. (2015). Graduating from social protection? Editorial introduction. *IDS Bulletin, 46*(2), 1–2. Retrieved August 14, 2021 at 11:00 am, from https://www.researchgate.net/publication/274645848_Graduating_from_Social_Protection_Editorial_Introduction.

Dharmadasa, H., Hashemi, S. M., Samaranayake, S., & Whitehead, L. (2015). *BRAC propel toolkit: An implementation guide to the ultra-poor graduation approach*. New York, USA. Retrieved December 28, 2020 at 15:00, from https://www.un.org/esa/socdev/egms/docs/2016/Poverty-SDGs/BRAC-PROPEL-Toolkit.pdf.

El-Zoghbi, M., Montesquiou, A. D., & Hashemi, S. (2009). *Creating pathways for the poorest: Early lessons on implementing the graduation model*. CGAP brief. Retrieved January 26, 2021 at 11:00 am, from https://www.cgap.org/sites/default/files/CGAP-Focus-Note-Reaching-the-Poorest-Lessons-from-the-Graduation-Model-Mar-2011.pdf.

Halder, S. R., & Mosley, P. (2004). Working with the ultra-poor: Learning from BRAC experiences. *Journal of International Development, 16*(3), 387–406.

Hashemi, S., & Umaira, W. (2011). *New pathways for the poorest: The graduation model from BRAC*. CSP Research Report 10, Brighton: Centre for Social Protection, IDS. Retrieved August 14, 2021 at 11:30 am, from https://gsdrc.org/document-library/new-pathways-for-the-poorest-the-graduation-model-from-brac/.

Hashemi, S. M., & De Montesquiou, A. (2011). Reaching the poorest: Lessons from the graduation model. *CGAP, Focus Note, 69*(1), 1–15.

Kidd, S., & Bailey-Athias, D. (2017). The effectiveness of the graduation approach: What does the evidence tell us. *Policy in Focus, 14*(2), 22–28.

Matin, I., & Hulme, D. (2003). Programs for the poorest: Learning from the IGVGD program in Bangladesh. *World Development, 31*(3), 647–665.

Roser, M., & Ortiz-Ospina, E. (2017). *Global extreme poverty*. Published online at OurWorldInData.org. Retrieved August 4, 2021 at 10:00, from https://ourworldindata.org/extreme-poverty.

Sabates-Wheeler, R., & Devereux, S. (2011). *Transforming livelihoods for resilient futures: How to facilitate graduation in social protection*. FAC Working Paper 23, Future Agricultures Consortium, Brighton.

Sabates-Wheeler, R., Sabates, R., & Devereux, S. (2018). Enabling graduation for whom? Identifying and explaining heterogeneity in livelihood trajectories post-cash transfer exposure. *Journal of International Development, 30*(7), 1071–1095.

Schelzig, K., & Rawal, P. (2020). *Testing the graduation approach in the Philippines*. Asian Development Bank Briefs No. 120, Manila, Philippines.

Sen, A. (1976). Poverty: An ordinal approach to measurement. *Econometrica, 44*(2), 219–231.

Sulaiman, M., Goldberg, N., Karlan, D., & de Montesquiou, A. (2016, December). Eliminating extreme poverty: Comparing the cost-effectiveness of livelihood, cash transfer, and graduation approaches. *In Forum. Washington, DC: CGAP. C.*

Tambe, S., Bhutia, N. T., Pradhan, S., & Basi, J. (2019). Coupling a ladder to the safety net: reinventing MGNREGA for asset creation. *Development in Practice, 29*(4), 514–524.

Van Domelen, J. (2007). Reaching the poor and vulnerable: Targeting strategies for social funds and other community-driven programs. In *Network HD* (pp. 1–52). World Bank.

World Bank Group. (2017). *A decade of rural transformation. Lessons learnt from the Bihar Rural Livelihoods Project—JEEViKA*. International Bank for Reconstruction and Development, World Bank. Retrieved June 16, 2021 at 11:00, from https://openknowledge.worldbank.org/handle/10986/29193?locale-attribute=en.

Chapter 9
Livelihood Resilience

Abstract Increasing impacts of disasters on human wellbeing have turned the spotlight on the urgent need to build resilience and safeguard livelihoods. The focus of this chapter is to understand the evolution of resilience thinking in terms of what constitutes resilience, how to measure it objectively, and further to apply it in development. Can it be thought of as a functional attribute of an ecosystem or livelihood system to maintain its functionality in response to change or a dynamic ability to resist collapse and even to flip and reorganize? At what scale and time frame does it operate? Is it specific to certain risks or is it generic? How to measure it objectively and how similar or dissimilar is it to other related concepts used in crisis contexts such as vulnerability and adaptation? Can there be good and bad resilience or is it always positive? How do we embed this concept in programmes aiming to build livelihood resilience? This chapter analyses these questions by taking a deep dive into resilience thinking, assimilating the diverse conceptualizations of resilience, differentiating resilience from related concepts, bringing together the various approaches to measuring livelihood resilience, and the lessons for livelihood practice.

Keywords Disasters · Shocks · Stress · Risks · Vulnerability · Absorptive · Adaptive · Transformative · Respond · Dignity · Wellbeing

9.1 Resilience Thinking Gaining Prominence

Livelihoods around the globe are increasingly caught in the transition of the environmental, economic, and social systems. More so in the global south, where livelihoods are in flux, from rural to the urban, farm to non-farm, from subsistence to market-led, from agriculture to manufacturing, from industry to service sector, from small holder farming to diversification, and vice versa. These transitions are either driven by enhanced capabilities and opportunities or can be seen as a coping mechanism in the face of an uncertain future. The DFID sustainable livelihoods approach (SLA) and livelihood trajectory studies provide insights into these dynamics. The SLA locates the household asset pentagon within a specific vulnerability context comprising of shocks, seasonality, and trends and the dynamic interaction between the two. The livelihood trajectory studies capture the dynamics of livelihoods in response to this vulnerability context and the larger development environment. In this context, one might wonder what is the value that resilience, which is a concept closer to engineering and natural sciences, can add to livelihood dynamics?

Livelihood trajectory studies use long-term panel data spanning across decades to analyse the livelihood pathways and transition matrices of households. These studies analyse the nature of the pathways out of and into poverty and the reasons behind the same. Over the last few decades, natural disasters, conflicts, political upheavals, economic crises, pandemics, and other disruptions are increasing in frequency and intensity. The response of households to this crisis context is becoming a critical and specialized area of study. Resilience in the context of livelihoods is a measure of the ability to

S. Tambe, *Teaching and Learning Rural Livelihoods*, Sustainable Development Goals Series,
https://doi.org/10.1007/978-3-030-90491-3_9

withstand shocks and stresses that are environmental, social, political, or economic. Hence, while livelihood trajectory studies are long term and retrospective, resilience thinking is proactive and has a future orientation, as it is about building capacities of households, communities, and institutions in the face of sudden shocks and long-term stresses. Off late resilience thinking is gaining prominence in livelihood studies especially in the domains of social protection, disaster risk reduction (DRR), and climate change adaptation (CCA) (Bahadur et al. (2010).

With rising exposure to shocks and stresses, the ability to achieve many of the SDG goals hinges on this key capacity. Resilience thinking is also embedded in the SDGs, specifically SDG 1 on ending poverty in all forms where target 1.5 indicates the need to build the resilience of the poor in the face of shocks and disasters. SDG 2 on ending hunger and achieving food security with target 2.4 emphasizing on the need to implement resilient agricultural practices, and SDG 13 that focusses on strengthening resilience and adaptive capacity to climate-related hazards and natural disasters. While resilience thinking is becoming popular in academics, even development agencies, governments, and practitioners are increasingly investing in strengthening resilience (Ngesa et al., 2020). Resilience building has also become one of the key objectives of humanitarian organizations that are investing in strengthening the disaster resilience agenda (DFID, 2011). Over the last decade, a growing number of bilateral and multilateral donor agencies have endorsed and embraced resilience thinking.

9.2 The Conceptualization of Resilience

Resilience as a concept in physical sciences and ecology emerged as early as the 1960s and was adapted into the humanitarian and development field in the 1990s (Béné et al., 2012; Ngesa et al., 2020). Resilience has been used in diverse settings such as resilience of materials in structural engineering, the resilience of ecosystems in the face of disturbance, psychological resilience of humans to mentally cope with trauma, and can have different meanings based on the context we are applying it in (Fig. 9.1). In this chapter, we focus on the resilience of human livelihoods in response to change. Adger (2000) defined resilience as 'the ability of groups or communities to cope with external stresses and disturbances as a result of social, political, and environmental change'. Walker et al. (2004) defined resilience as 'the capacity of a system to absorb disturbance and reorganize while undergoing change to still retain essentially the same function, structure, identity, and feedbacks'. Marschke and Berkes (2006) talk of stress and shock and how households respond and continuously adjust to change. Resilience is future oriented and offers a lens with which to explore the crisis context and to understand livelihood dynamics. FAO describes resilience as 'The ability to prevent disasters and crises as well as to anticipate, absorb, accommodate, or recover from them in a timely, efficient, and sustainable manner'. Fraser et al. (2011) conceptualized a three-dimensional vulnerability framework in coupled socio-ecological systems. This framework has three components, namely the resilience of the ecosystem, the resilience of the livelihoods, and the institutional capability to mobilize effective relief during a crisis.

The aforementioned definitions of resilience drawn from different perspectives, put at their centre the concepts of 'absorbing', 'accommodating', and 'responding to' disturbance, but focus on the resilience of community or system (Pain & Levine, 2012). Tanner et al. (2015) state that resilience in policymaking has often been based on the ability of systems to bounce back to normality, drawing on engineering concepts. This implies the return of the functions of an individual, household, community, or ecosystem to previous conditions, with as little damage and disruption as possible following shocks and stresses. They argue that looking at resilience as a purely scientific and technical approach undermines the role that people can play in driving positive transformation. Hence, they proposed a livelihood perspective to understanding resilience by placing greater emphasis on human needs and their agency, thereby placing them at the centre of the resilience debate.

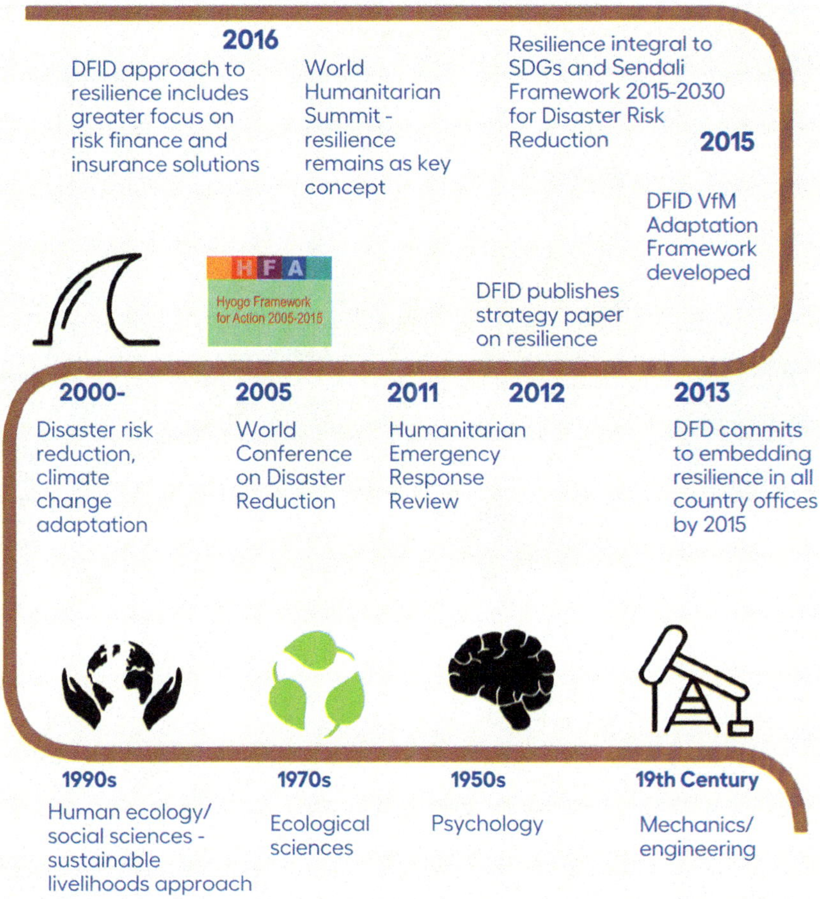

Fig. 9.1 The rise of resilience across disciplines with different meanings. (Source DFID, 2016)

Resilience is also conceptualized as an outcome by building on the four elements of context, disturbance, capacity, and resilience (Fig. 9.2) DFID (2011). Context and disturbance are key in understanding resilience. It is, therefore, crucial to ask two questions, 'resilience of what' and 'resilience to what' (Smith & Stirling, 2010). The 'context' can be a household, community, institution, or also of a socio-economic system. The 'disturbance' can take the form of sudden events such as disasters, disease, violence, conflicts, accidents, etc., while the stresses are long-term trends such as natural degradation, urbanization, demographic changes, economic decline, and the like. The 'capacity' is the ability to deal with shocks and stress and is based on the levels of exposure, sensitivity, and adaptive capacities. The 'reaction' to a shock or system could follow four pathways—'bounce back better', 'bounce back', 'recover but worse than before', or 'collapse'.

Although resilience has been conceptualized as an 'outcome' or as a 'target' that can be aspired for and tracked, there is a growing consensus now that a more useful way to understand it is as an 'ability' or 'capacity' to deal with adverse changes and shocks (Béné et al., 2012). Ngesa et al. (2020) conceptualize resilience as 'a capacity that is driven by three interrelated capacities, namely, absorptive capacity, adaptive capacity, and transformative capacity. Absorptive capacity refers to the social and technical skills and strategies to reduce exposure to shocks to minimize the negative impact of hazards. Absorptive capacity can be signified by stability despite the occurrence of shocks and hazards. Adaptive capacity refers to the ability to make proactive, informed choices and changes in livelihood

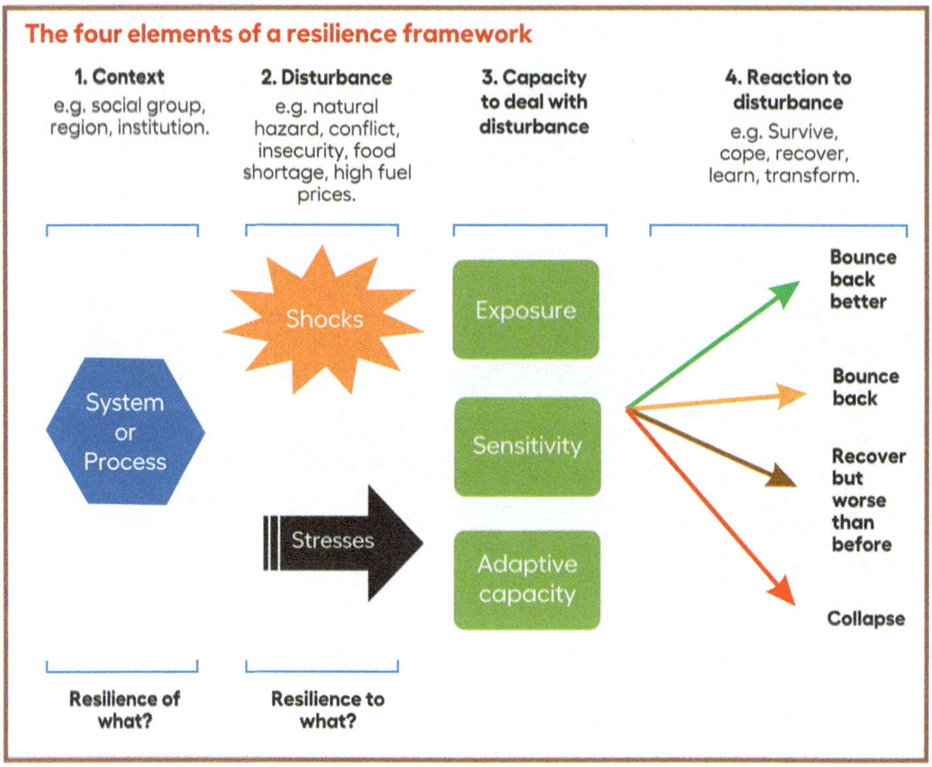

Fig. 9.2 Resilience conceptualized as an outcome. (Source DFID, 2011)

and other strategies in response to longer-term social, economic, and environmental change. Adaptive capacity can be signified by the ability to adjust to shocks and hazards. Transformative capacity refers to the governance mechanisms, leadership, policies, and infrastructure that provide an enabling environment for change. The transformative aspect addresses the root causes of vulnerability to shocks'.

Béné et al. (2012) designed the 3D resilience framework highlighting that resilience is an outcome of three capacities, each of which results in differential outcomes (Fig. 9.3). The absorptive capacity results in persistence, adaptive capacity in incremental adjustment, and transformative capacity in transformational response. Understood in a linear way, the framework promotes resistance in the period of small disturbance, adaption in the face of moderate disturbance, and transformability when conditions are unsustainable. Also, resilience exists at multiple scales: individual, household, community, society, etc. They then combine the 3D resilience framework with the Protection-Prevention-Promotion-Transformation (3PT) framework. 'Protection' measures include social policies and instruments aimed at protecting destitute and marginalized households such as children, orphans, elderly, disabled, etc. by providing social welfare measures such as pension, food transfer, cash stipends, and the like. 'Preventive' schemes are the safety net programmes that prevent the households from falling deeper into poverty, for instance, unemployment programmes, food subsidies, etc. 'Promotion' programmes aim to enhance livelihoods by building capacity, skill development, strengthening the asset mix, and providing access to resources. 'Transformative' measures focus on structural issues by addressing concerns of social justice and exclusion through modifications in policies, laws, programmes, and budgetary allocations. Some of the programmes and measures may have overlap and simultaneously 'prevent' deprivation while also 'promote' incomes (Béné et al., 2012). The 3PT framework and the 3D resilience framework can be merged into a two-dimensional assessment matrix (Fig. 9.4). This matrix provides an analytical framework to evaluate how development programmes

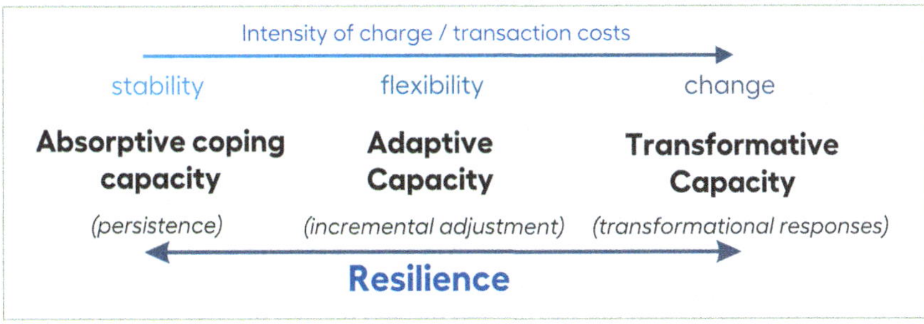

Fig. 9.3 The 3D Resilience Framework conceptualized as a capacity. (Source: Béné et al., 2012)

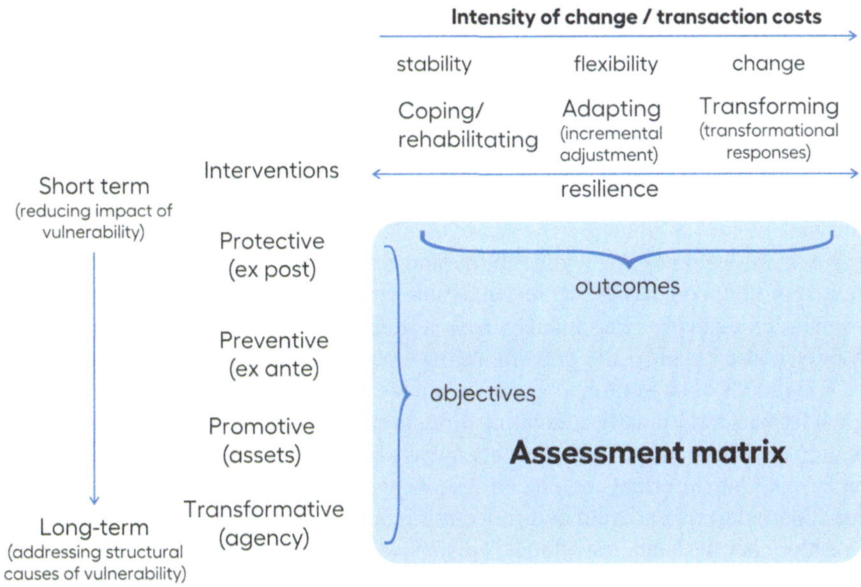

Fig. 9.4 The 3PT-3D analytical framework. (Source: Béné et al., 2012)

contribute to strengthening resilience and through which dimension, i.e. coping, adapting, or transforming (Béné et al., 2012).

Tanner et al. (2017) caution against using resilience as a guiding narrative or a normative concept and being used as a goal or a development objective as it is a neutral characteristic that is inherently neither good or bad. Like governance, which can be good and bad, we need to talk of bad resilience as well (Béné et al., 2012). An apt example of bad resilience is an authoritarian regime that manages to stay in power through torture and repression despite attempts to bring in more democracy. Berkhout (2008) highlights the evil side of resilience and points out that being resilient may not always be an ideal outcome, and we may want to do away with systems that endorse inequality, discrimination, environmental degradation, or a carbon-intensive development. Hence, to understand resilience, we reiterate the importance of asking the two key questions 'resilience of what' and 'resilience to what', i.e. what precisely is being made resilient and in the face of which specific dynamics and whether this resilience is good or bad (Smith & Stirling, 2010). Béné et al. (2012) also reveal that poverty alleviation and resilience building are two different concepts as typically households can be very poor and very resilient at the same time and non-poor households may pursue non resilient livelihoods.

Therefore, resilience should not be looked upon as a pathway out of poverty and as a replacement for poverty alleviation.

We note that over the years the definition of resilience has evolved and broadened from the initial framing as the ability of a system to bounce back, into an elaborate concept that embraces not only bounce back but also to adapt and to transform (Béné et al., 2012). Within social sciences, where owing to its engineering roots the initial focus was on the resilience of the system and resilience of communities. It is only over the last few years that household livelihood resilience has started gaining prominence (Quandt, 2018). Also, resilience should not be looked upon as a goal in itself, as there can be numerous instances and scenarios when strengthening resilience may be detrimental to human wellbeing and vice versa (Béné et al., 2012).

9.3 Differentiating Resilience

While the concepts of poverty, vulnerability, adaptation, and resilience are used in similar contexts, it is important to understand the intricacies and ask questions on how they differ. Are the poor more vulnerable or less resilient in the face of disasters and risks? Is adaptation to shocks and stresses always positive or it can have negative outcomes as well? Is resilience an innate ability to retain functionality in the face of change or is it a dynamic ability that can be enhanced? Is resilience a capacity or is it an outcome? Are vulnerability and resilience the opposite ends of the same axis? While intuitively the poor may be more vulnerable to risks and disasters, this may not always be borne by empirical evidence. Earthquakes pose a greater risk to middle class families residing in stone masonry houses than to the poor living in wood and bamboo houses. Also, droughts may impact the livelihoods of large farmers growing water-intensive sugarcane much more than the small and marginal farmers traditionally cultivating drought-resistant millets. Tanner et al. (2015) point out that adaptation to shocks and stresses may not always be positive as households may be forced from traditional farming into informal and precarious jobs without decent work conditions or pay. Kothari and Hulme (2004) describe a livelihood trajectory in a case from Bangladesh, where following the untimely death of her husband, the widow is dispossessed of her agricultural land which is her only source of income. She finds it difficult to find work and starts working part time as a domestic help while also relying on charity and begging to meet her daily needs. This case highlights that adaptation may not always result in a positive livelihood transformation. This is what differentiates livelihood resilience which is framed not as an outcome, but as the building of capacities and the ability of the most vulnerable to respond to shocks and stresses in ways without compromising on human wellbeing or dignity.

Pain and Levine (2012) argue that while the term 'vulnerability' refers specifically to susceptibility to harm from a specified threat, but it is misleading to treat resilience and vulnerability as polar opposites, i.e. if a community is more vulnerable to a shock, then it will be less resilient and vice versa (Fraser et al., 2011). A fishery community, for example, can be seen as highly vulnerable due to high illiteracy, low-income levels, and weak safety nets, but it may be highly resilient due to abundant fish stock, strong social networks, and rich cultural capital (Thanh et al., 2020). Béné et al. (2012) point out that though Japan and Peru were hit by earthquakes of similar magnitude, the death toll in Japan was 63 as compared to 2900 in Peru. Also, in 1998, Hurricane Georges resulted in 589 human casualties in the Dominican Republic and Haiti, while only 6 were killed in Cuba. They reason that though the vulnerabilities did not differ by much in these cases, the difference in outcomes was a result of the preparedness interventions building resilience in Japan and Cuba.

9.4 Measuring Household Livelihood Resilience

Resilience is conceptualized as a capacity or ability and is hence not easy to measure. Over the last few years frameworks to measure household livelihood resilience by building on surrogate indicators have been developed. We describe two such methods, namely the Household Livelihood Resilience Approach (HLRA) and the Resilience Index Measurement and Analysis (RIMA).

Quandt (2018) developed the Household Livelihood Resilience Approach (HLRA) for measuring household-level resilience. Resilience being a capacity or an ability, it is difficult to measure it directly and hence this approach uses quantifiable surrogates. It builds on the theoretical framework of the DFID SLA as the analytical framework. A total of 25 quantitative indicators were used to build the five capitals and then aggregated to construct the composite asset index. The visualization was done with the help of spider graphs to enable comparative analysis between different groups. The HLRA helps in analysing, visualizing, and interpreting the results and in identifying the strategy and interventions needed for building resilience (Quandt, 2018). However, being context specific and dependent on intensive household-level fieldwork, it is challenging to link it to policymaking.

The RIMA methodology is used to calculate household resilience capacity based on four key pillars, these being access to basic services (ABS), assets (AST), social safety nets (SSN), and adaptive capacity (AC) (Ngesa et al., 2020). Observed household-level variables are used to construct these sub-components that are aggregated to develop the composite Resilience Capacity Index (RCI) for the measurement of resilience at the household level. The estimated RCI is then scaled in the range of 0–100 with a higher score indicating higher household-level resilience. Sarker et al. (2020) explore the livelihood resilience of vulnerable riverine island dwellers in Bangladesh by employing the Resilience Index Measurement and Analysis II (RIMA II) model developed by Food and Agriculture Organization (FAO) for measuring resilience at the household level (FAO, 2016). According to this model, the resilience outcome of a household is a function of the natural disaster probability, probability of exposure of the household, and resilience of the household. The resilience is again a function of the three dimensions of absorptive, adaptive, and transformative capacity which is assessed by building on the SLA to develop indicators and aggregating them (Figs. 9.5 and 9.6). It allows the indirect measurement of resilience using proxy parameters that can be easily collected through household surveys and adapted to the local context. Furthermore, from the RIMA results, inferences can be made on the main drivers of resilience, therefore allowing for the tailoring of resilience building interventions. Comparisons can be made of the results spatially and temporally, enabling the assessment of the effectiveness of different resilience building interventions.

9.5 Lessons from Research and Practice

Resilience building has also become one of the key objectives of humanitarian organizations that are investing in strengthening the disaster resilience agenda by financing, advocacy, networks, knowledge, and integration DFID (2011). However, what does it mean for a particular programme to be resilience-driven, what would the programme components look like? We share some of the lessons from research and practice (Boxes 9.1 and 9.2).

Box 9.1 Resilience Building in Practice (Source DFID, 2011)

DFID Ethiopia runs a Productive Safety Net Programme that covers 7.8 million vulnerable people provisioning regular cash and food transfers. A new risk financing mechanism allows the programme to expand in times of shock by increasing the time over which an individual receives transfers or add more people to the programme. This mechanism has helped in protecting the asset base of the households in times of shock. DFID Bangladesh strengthens resilience to climate change by introducing early warning systems, raised plinths for villages to protect them from flooding, renovated embankments and roads, multipurpose cyclone shelters, and climate-resilience crops. DFID Africa established the Africa Risk Capacity, which will establish a pan-African disaster risk pool for food security. The initiative will provide participating countries with funds to manage risk and respond to extreme weather events.

Source: DFID (2011)

Box 9.2 The R4 Rural Resilience Initiative

Farming continues to be the mainstay of more than 1.2 billion people in the developing world, with climate-related shocks posing a real threat to their wellbeing and food security. The R4 Rural Resilience initiative was set up as a strategic collaboration between the World Food Program and Oxfam America. It was launched in 2011, to build the resilience of vulnerable rural families in several African nations to increase their long-term food security and livelihoods improvement by managing climate-related risks. The initiative combines four risk management strategies:

- Risk reduction: to improve resource management through asset creation
- Risk transfer: to provide access to insurance and help reduce vulnerabilities over time
- Prudent risk taking: to promote livelihood diversification and access to microcredits
- Risk reserves: to help build a stronger financial base for the communities

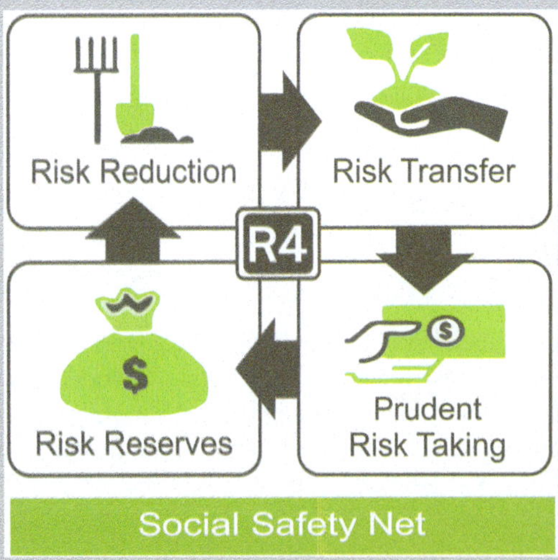

(continued)

Box 9.2 (continued)

In 2020, despite the challenges posed by COVID-19, the R4 Rural Resilience Initiative reached nearly 180,000 farming households (55% women), benefitting approximately 900,000 people in Bangladesh, Burkina Faso, Ethiopia, Kenya, Madagascar, Malawi, Mozambique, Senegal, Zambia, and Zimbabwe. Impact evaluation studies have indicated that the initiative is helping improve resilience of these communities in several ways including gender equality, asset building, insurance pay-outs for weather-related losses, and livelihood diversification.

Source: World Food Programme (2020)

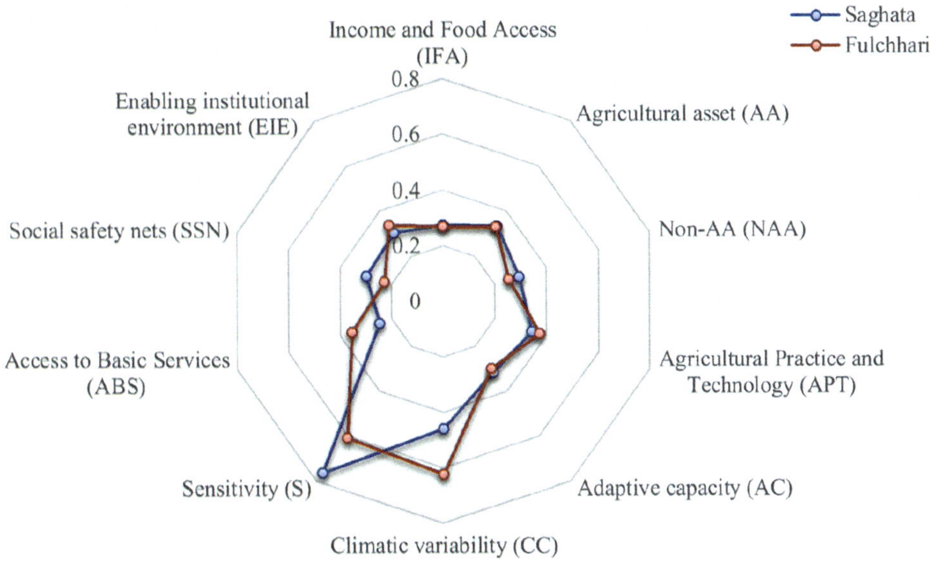

Fig. 9.5 Spider diagram of major components of the livelihood resilience index of riverine island dwellers in the two villages of Saghata and Fulchhari of Bangladesh. (Source: Sarker et al., 2020)

Fig. 9.6 Resilience triangle of major components of the vulnerability of riverine island dwellers in the two villages of Saghata and Fulchhari of Bangladesh. (Source: Sarker et al., 2020)

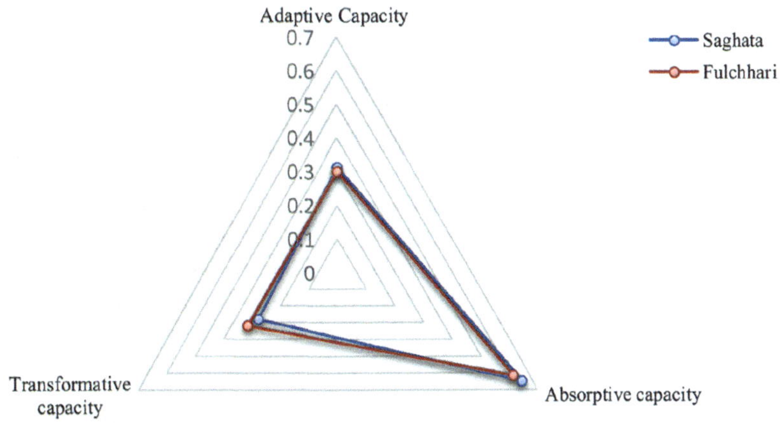

Ngesa et al. (2020) carried out RIMA (Resilience Index Measurement and Analysis) analysis in the Horn of Africa in the three countries of Kenya, Uganda, and Ethiopia. They found that in Kenya, asset ownership and adaptive capacity are crucial to resilience and drought is the key shock. While livestock adaptation boosts the resilience of households the most compared to crop and off-farm adaptation, however, they are effective for households who are already in higher resilience categories. In Uganda as well, asset ownership and adaptive capacity are crucial in resilience building, while crop diseases, pests, and water shortage are the key shocks. Off-farm adaptation practices boost the resilience of the households the most compared to crop, livestock, or off-farm activities. In Ethiopia assets, access to basic services, and social safety nets contribute the most to household resilience. High food prices and water shortage are the main shocks while livestock adaptation boosts household resilience the most as compared to crop and off-farm adaptation practices. Across all the three countries, women-headed households had generally lower resilience index scores compared to the other male-headed households. These case studies indicate how in different regions, the resilience capacity is driven by different factors, is undermined by different types of shocks, and how the adaptation practices vary based on the locality. These studies, therefore, help in building an argument for the need for resilience programming to not take a blanket approach but rather an approach that is tailored to the specific needs and capacities of households and communities.

Gautam (2017) studied seasonal migration and its role in livelihood resilience in the high Himalayas of north-western Nepal. He found that it was not climate change but structural poverty that was the main driver of migration, with men taking up low-paying unskilled work in the adjacent regions in India and make only modest earnings which will not help in significantly building resilience. Kaur et al. (2017) studied the role of public work programmes in building resilience to climate-induced droughts in the Himalayan state of Sikkim in India. They show that the public work interventions can lead to expansion of the asset pentagon, which in turn can help strengthen household resilience to climate change at three levels, by enabling them to absorb, adapt, or transform to manage risks and opportunities. A study by Sarker et al. (2020) described in an earlier sub-section in the context of measuring household livelihood resilience of riverine island dwellers in Bangladesh used the RIMA model to measure the adaptive, absorptive, and transformative capacity. They found that the riverine island dwellers have a poor level of resilience which made them unable to withstand natural disasters. Those residing near the mainland area displayed relatively more resilience than those in the interior areas. The main limiting factors impacting livelihood resilience include access to food, income, health facilities, assets, and technology adoption.

9.6 Concluding Remarks

The conceptualization of resilience has evolved from engineering to psychology to ecological sciences to socio-ecological sciences, and finally to household-level livelihoods. Various models have been now developed to measure household resilience and quantify it as an index or score. These models not only measure resilience but also provide guidance on designing resilience as a three-dimensional ability (adaptive, absorptive, transformative) and embedding it in programmes that have four dimensions (protection, prevention, promotion, and transformation). These two components can be integrated to form a matrix to assess the impact of various components of programmes on the various dimensions of resilience. One danger that resilience poses to development is if it becomes a non-negotiable with policymakers and donors insisting on its inclusion in every project they support. We need to be careful not to treat resilience as a normative concept, as it can be either good or bad just like governance. Also, it is crucial to acknowledge that resilience must not be looked upon as a tool to alleviate poverty as the poor too, in many cases may already be resilient. We opine that in the decades to come rising crisis contexts will challenge the sustainability of the achievements made during the

MDG and SDG period. As we confront unprecedented challenges on a planetary scale in the Anthropocene, development practice will need to refocus on building resilience so that human wellbeing is sustained.

Exercises

1. Davies et al. (2013) advocate the integration of social protection (SP), disaster risk reduction (DRR), and climate change adaptation (CCA) programmes in the agriculture sector under the adaptive social protection (ASP) framework. They argue that these programmes have the same objective of vulnerability reduction but are often implemented in isolation by different ministries. In the context of sustainable rural livelihoods, discuss the potential advantages and flip side of this approach? https://www.researchgate.net/publication/256042033_Promoting_Resilient_Livelihoods_Through_Adaptive_Social_Protection_Lessons_from_124_Programmes_in_South_Asia
2. Critically analyse the five frameworks of vulnerability espoused by Chambers and Conway (1992), Carney (2002), Fraser et al. (2011), and Jeans et al. (2016). These studies can be accessed from the details provided in the reference section of this chapter.
3. Sallu et al. (2010) draw on the concepts of livelihood trajectories and resilience to assist in the exploration of vulnerability in the drylands of Botswana. Discuss the factors leading to resilience and vulnerability of the households and the broad steps needed towards strengthening livelihood resilience. https://www.jstor.org/stable/26268197
4. Quandt (2018) constructs an innovative Household Livelihood Resilience Approach (HLRA) for measuring livelihood resilience and elucidates its five strengths. Analyse the strengths and weaknesses of this model with justification. https://doi.org/10.1016/j.worlddev.2018.02.024
5. Quandt et al. (2019) studied the role that agroforestry plays in improving the overall quality of life of households in Kenya. Describe how the study builds on the sustainable livelihoods framework to construct the overall livelihood score. https://doi.org/10.1080/17565529.2018.1447903
6. Kaur et al. (2017) study the vulnerability to climate change in the Himalayan state of Sikkim, India, and the role of a public works programme in building resilience and assess the pathways to resilience and the resultant outcomes. Analyse the integration of the DFID sustainable livelihoods approach with the resilience framework and the conclusions of this study. https://pubs.iied.org/pdfs/10188IIED.pdf
7. Adger et al. (2002) study the relationship between demographic change, social resilience, and sustainable development in coastal Vietnam. Analyse the impacts of economic inequality, diversification, and increase in income levels on resilience. https://doi.org/10.1579/0044-7447-31.4.358
8. Gautam (2017) studied the drivers of seasonal migration and livelihood resilience in Nepal and the factors affecting the farmer's decision to migrate. He found that rather than climate change impacts, structural poverty was the root cause of migration. Analyse his conclusion that migration is unlikely to make a significant contribution to building livelihood resilience in the context of climate change in remote Himalayan farming communities. https://doi.org/10.1659/MRD-JOURNAL-D-17-00035.1
9. Ngesa et al. (2020) use the RIMA II (Resilience Index Measurement and Analysis II) as a primary methodology to guide agricultural climate change adaptation and resilience building investments in selected communities in three East African countries. Analyse the benefits and limitations of the RIMA methodology for policymakers, agricultural development practitioners, and rural livelihood professionals. https://doi.org/10.1007/978-3-030-22759-3_67-1
10. Sarker et al. (2020) study the livelihood resilience of riverine island (*char*) dwellers in the face of natural disasters in Bangladesh. They collect data using face-to-face structured interviews and focus group discussions to construct the Livelihood Resilience Index and Resilience triangle of the char dwellers. Familiarize yourself with the data collection, data analysis, and construction of

these composite indicators. Examine the recommended development interventions in detail and list down a few hypothetical interventions not indicated in the study that can increase the livelihood resilience of the riverine island dwellers. https://doi.org/10.1016/j.landusepol.2020.104599

References

Adger, W. N. (2000). Social and ecological resilience: Are they related? *Progress in Human Geography, 24*(3), 347–364.

Adger, W. N., Kelly, P. M., Winkels, A., Huy, L. Q., & Locke, C. (2002). Migration, remittances, livelihood trajectories, and social resilience. *Ambio: A Journal of the Human Environment, 31*(4), 358–366.

Bahadur, A. V., Ibrahim, M., & Tanner, T. (2010). *The resilience renaissance? Unpacking of resilience for tackling climate change and disasters*. Strengthening Climate Resilience Discussion Paper No. 1, IDS. Brighton. Retrieved January 14, 2021 at 9:00, from https://opendocs.ids.ac.uk/opendocs/handle/20.500.12413/2368.

Béné, C., Wood, R. G., Newsham, A., & Davies, M. (2012). Resilience: New utopia or new tyranny? Reflection about the potentials and limits of the concept of resilience in relation to vulnerability reduction programmes. *IDS Working Papers, 2012*(405), 1–61.

Berkhout, F. (2008). Order in socio-technical systems: The dark side of resilience. In M. Leach (Ed.), *Re-framing resilience: A symposium report. STEPS Working Paper 13*. Institute of Development Studies, STEPS Centre.

Carney, D. (2002). *Sustainable livelihoods approaches: Progress and possibilities for change*. Department for International Development.

Chambers, R., & Conway, G. (1992). *Sustainable rural livelihoods: Practical concepts for the 21st century*. IDS Discussion Paper 296, Institute of Development Studies, Brighton.

Davies, M., Béné, C., Arnall, A., Tanner, T., Newsham, A., & Coirolo, C. (2013). Promoting resilient livelihoods through adaptive social protection: Lessons from 124 programmes in South Asia. *Development Policy Review, 31*(1), 27–58.

DFID (Department for International Development). (2011). *Defining disaster resilience: A DFID approach paper* (pp. 1–20). UK. Retrieved from https://reliefweb.int/sites/reliefweb.int/files/resources/Defining%20Disaster%20Resilience%20A%20DFID%20Approach%20Paper%20%28new%20cover%29.pdf.

DFID (Department for International Development). (2016). *What is resilience? DFID approach paper* (pp. 1–65) Retrieved from: https://assets.publishing.service.gov.uk/media/57a08955ed915d3cfd0001c8/EoD_Topic_Guide_What_is_Resilience_May_2016.pdf

FAO. (2016). *Analysing resilience for better targeting and action*. Food and Agriculture Organization of the United Nations.

Fraser, E. D., Dougill, A. J., Hubacek, K., Quinn, C. H., Sendzimir, J., & Termansen, M. (2011). Assessing vulnerability to climate change in dryland livelihood systems: Conceptual challenges and interdisciplinary solutions. *Ecology and Society, 16*(3).

Gautam, Y. (2017). Seasonal migration and livelihood resilience in the face of climate change in Nepal. *Mountain Research and Development, 37*(4), 436–445.

Jeans, H., Thomas, S., & Castillo, G. (2016). The future is a choice: The Oxfam framework and guidance for resilient development. .

Kaur, N., Steinbach, D., Agrawal, A., Manuel, C., Saigal, S., Panjiyar, A., Shakya, C., & Norton, A. (2017). *Building resilience to climate change: MGNREGS and climate-induced droughts in Sikkim*. IIED Issue Paper, IIED, London.

Kothari, U., & Hulme, D. (2004). *Narratives, stories and tales: Understanding poverty dynamics through life histories*. GPRG-WPS-011. ESRC Global Policy Research Group, Manchester, 41 pp.

Marschke, M. J., & Berkes, F. (2006). Exploring strategies that build livelihood resilience: A case from Cambodia. *Ecology and Society, 11*(1).

Ngesa, O., Grey, S., Duveskog, D., & Atieno, I. (2020). Characteristics and determinants of the resilience of smallholder farmers: Lessons from application of the RIMA II methodology in eastern Africa. In W. Leal Filho, J. Luetz, & D. Ayal (Eds.), *Handbook of climate change management*. Springer.

Pain, A., & Levine, S. (2012). *A conceptual analysis of livelihoods and resilience: Addressing the insecurity of agency*. ODI.

Quandt, A. (2018). Measuring livelihood resilience: The household livelihood resilience approach (HLRA). *World Development, 107*, 253–263.

Quandt, A., Neufeldt, H., & McCabe, J. T. (2019). Building livelihood resilience: What role does agroforestry play? *Climate and Development, 11*(6), 485–500.

Sallu, S. M., Twyman, C., & Stringer, L. C. (2010). Resilient or vulnerable livelihoods? Assessing livelihood dynamics and trajectories in rural Botswana. *Ecology and Society, 15*(4).

Sarker, M. N. I., Wu, M., Alam, G. M., & Shouse, R. C. (2020). Livelihood resilience of riverine island dwellers in the face of natural disasters: Empirical evidence from Bangladesh. *Land Use Policy, 95*, 104599.
Smith, A., & Stirling, A. (2010). The politics of social-ecological resilience and sustainable socio-technical transitions. *Ecology and Society, 15*(1).
Tanner, T., Bahadur, A., & Moench, M. (2017). *Challenges for resilience policy and practice*. Working Paper 519. Overseas Development Institute, London.
Tanner, T., Lewis, D., Wrathall, D., Bronen, R., Cradock-Henry, N., Huq, S., et al. (2015). Livelihood resilience in the face of climate change. *Nature Climate Change, 5*(1), 23–26.
Thanh, H. T., Tschakert, P., & Hipsey, M. R. (2020). Moving up or going under? Differential livelihood trajectories in coastal communities in Vietnam. *World Development, 138*, 105219.
Walker, B., Holling, C. S., Carpenter, S. R., & Kinzig, A. (2004). Resilience, adaptability and transformability in social–ecological systems. *Ecology and Society, 9*(2), 5.
World Food Programme. (2020). *Annual Report*. Rome. Retrieved August 10, 2021 at 11:40 am, from https://www.wfp.org/publications/r4-rural-resilience-initiative-2020-annual-report.

Chapter 10
Synthesis of the Approaches

Abstract The purpose of this chapter is to develop a deeper understanding of the comparative strengths and limitations of the six livelihood approaches, namely sustainable livelihoods approach, commons and livelihoods, livelihood trajectories, rights-based approach, graduation approach, and resilience framework. We assess these livelihood approaches against a set of parameters such as rationale, applicability, strengths, and weakness to develop a deeper understanding of their pros and cons. The purpose is to be able to analyse which approach is best suited in a given context or situation and what value it can add. We also explore if these approaches can inform each other, and the possibilities of fusing them to develop a more complete understanding of rural livelihoods. We also describe the comprehensive social protection typology comprising of protection, prevention, promotion, and transformation (3PT), and use this lens to analyse the comprehensiveness of the various livelihood approaches. We opine that a nuanced understanding of these six livelihood approaches and their suitable integration can play a key role in enhancing their applicability in research, policy, and practice.

Keywords Strengths · Weakness · Protection · Prevention · Promotion · Transformation · Integration · Convergent sequential · Convergent parallel · Niche area · Functionality · Ease of application

10.1 Synopsis of the Livelihood Approaches

We undertake a review of these six livelihood approaches, namely sustainable livelihoods approach, commons and livelihoods, livelihood trajectories, rights-based approach, graduation approach, and the resilience framework (Table 10.1). These approaches when applied at the village level can unearth different aspects of complex livelihoods thereby contributing to our understanding of the larger picture.

10.1.1 The DFID Sustainable Livelihoods Approach

The DFID SLA can assist in an integrated livelihood assessment at the household level in a comprehensive manner by covering the three components of vulnerability context, asset mix, the prevailing policies, institutions and processes (PIP), and their inter-linkages (Scoones, 1998; DFID, 1999). It highlights the importance of the asset mix covering human, social, physical, financial, and natural capitals and thus moves beyond the focus on only monetary assets. The dynamic nature of poverty and livelihoods can be captured by monitoring the livelihood outcomes which in turn cause an expansion or contraction of the asset pentagon. This approach is implemented at the household level and is thus highly contextualized to be able to inform either project strategy or development policy. While this approach provides an in-depth understanding of the theoretical or conceptual part, applying it using

S. Tambe, *Teaching and Learning Rural Livelihoods*, Sustainable Development Goals Series,
https://doi.org/10.1007/978-3-030-90491-3_10

Table 10.1 Synopsis of the various livelihood development approaches

Livelihood approach	Key points
DFID SLA	Looks at household assets in five dimensions, factors in vulnerability context into livelihood planning, links macro with micro by highlighting the role played by policies, institutions, and processes. Livelihood dynamics can be represented through livelihood outcomes and feedback loops, resulting in an expansion or contraction of the asset pentagon. The livelihood strategy element is not detailed, but expected livelihood outcomes are spelled out
Commons and livelihoods	Highlights the importance of commons such as forests, fishery, pastures, irrigation, etc. in sustaining the livelihoods of the poorest. Common pool resources provide a safety net for the poorest by provisioning food, medicine, and cash income. Common property regimes ensure local enforcement to regulate access and withdrawal. This approach is focussed more on the commons and their governance and the role they play in sustaining local livelihoods
Livelihood trajectory	Analyses the past and present to delineate the livelihood pathways and categorize the trajectory. A quantitative transition matrix helps in providing a macro assessment of the poverty dynamics at the village level. Qualitative household interviews cover both the macro context and the household-level livelihood journey and thus help in ascertaining the causal links. It provides insights on what is working and what is not, and is thus useful to synthesize lessons for the future
Rights-based approach	The rights-based approach is specially tailored for addressing endemic, structural poverty and can target the root causes of the generation and perpetuation of poverty and deprivation. It is based on three key principles, namely participation, inclusion, non-discrimination, equality, and accountability. Development is then looked upon not as charity, but as an effort to fulfil rights, with local communities transforming from passive recipients to active right holders
Graduation approach	Especially tailor made for the poorest who often get left behind in mainstream development programmes. Step-by-step approach combining protection, prevention and promotion measures, comprising of consumption support, savings, skill development, productive asset transfer, microfinance, and life skills coaching
Resilience framework	With growing disasters and risks, there is an emerging concern regarding the sustainability of the livelihood outcomes achieved. In this regard, embedding resilience which is defined as an ability or capacity having three dimensions, namely absorptive, adaptive, and transformative is gaining prominence. Building resilience in development will be the key in the Anthropocene to sustain human wellbeing

practical methods and tools is the challenge (Quandt, 2018). The constituent elements of this approach such as vulnerability context, asset mix, and PIP are abstract, complex, and dynamic, thereby making the assessment of these components expert-driven (Jones & Tanner, 2015). Given that the assessment is household-centric, several livelihood experts will be needed to complete a time-bound livelihood study at a scale that will be able to inform project planning or policymaking. This approach is ideally suited for research projects, where academicians take up field-based studies covering a small area in an intensive manner. Also, in externally aided projects where there is provision for hiring consultants and early career researchers, these kinds of intensive and complex studies are feasible. In regular government-funded programmes, fewer resources are available for planning, as the prime focus is on implementation using in-house staff. Hence, in public programmes, often it is not feasible to take up these intensive, deep-dive studies, owing to a scarcity of human and financial resources.

10.1.2 Commons and Livelihoods

The collective governance of the commons framework turns the spotlight on the common pool resources (CPRs) such as grasslands, fisheries, forests, reservoirs, etc., and their potential in provisioning livelihoods for the poorest. Often while planning for the poor, the contribution of these commons to their livelihoods is overlooked. As the poor lack personal assets, these CPRs play the role of a safety net by provisioning famine food, medicines, and cash income (Wunder, 2011). However, as these are open access resources they can be easily over-consumed and the CPR can thus get degraded. Often, common property regimes exist, where user access is regulated with local enforcement (Ostrom, 1990). However, CPRs under weak local enforcement, face the risk of 'tragedy of the commons' and other challenges such as elite capture, curtailed access of weaker sections, etc. that can undermine their contribution to the livelihoods of the poor. Hence, it is vital to acknowledge the commons, restore their health, fortify their governance, and strengthen their access by the poor.

10.1.3 Livelihood Trajectories

Livelihoods are dynamic as households respond to shocks, stresses, and opportunities and consequently escape or fall back into poverty. The livelihood trajectory approach captures three vital aspects of poverty: how it is perceived by the local community, how it has changed and the drivers of change (Kothari & Hulme, 2004). By assessing the past and the present, it can guide livelihood development by informing what is working and what is not in a particular context. This approach uses a mix of quantitative and qualitative methods to answer these questions and creates a typology of the livelihood trajectories (Dorward et al., 2009). The quantitative data is collected in group exercise mode and helps in discerning the percentage of households who have escaped or fallen back into poverty. One of the highlights of this approach is generating the transition matrix with household-level data in quick time. This matrix provides insights on the livelihood trajectories operating at the village level, by quantifying the proportion of households who have escaped poverty, fallen back, remained poor, or remained non-poor. The quantitative part is followed by the qualitative assessment, where the life story of the poor is documented through household interviews to ascertain the causal links. This approach can help discern the drivers of poverty and thus has the potential to influence decision makers with these real and impactful stories. The livelihood trajectory assessment seamlessly feeds into the design of the future livelihood strategy by providing guidance on two components, one to block descent and the other to accelerate ascent (Krishna, 2006). Other than planning, this approach can also be used in monitoring and evaluation of livelihood projects, as the livelihood trajectory of households can be tracked over time and the attribution of the interventions ascertained. Livelihood trajectories provide important insights and need to be brought to the policy table more frequently.

10.1.4 Rights-Based Approach

The notions of the welfare state that dominated twentieth century development discourse, described poverty as a lack of capability, assets, or public good (Offenheiser & Holcombe, 2003). This model essentially surmised that delivery of development and public goods by the state would result in poverty alleviation (de Silva, 2013). After following this approach for several decades, it was comprehended that this form of development will not be able to address the deeper, structural causes of poverty linked to social justice (Offenheiser & Holcombe, 2003). This gave rise to the emergence of the rights-based approach where the focus is on structural barriers that impede communities from exercising rights, building capabilities, and having the freedom to choose. It assumes that poor people have dignity, aspirations, and ambition and that their initiative is being blocked and frustrated by persistent systemic challenges, such as apartheid, discrimination, biased lending policies, and non-functioning state social service delivery systems (Offenheiser & Holcombe, 2003). Executing development projects using this approach would need an in-depth analysis of the distribution of power, by identifying groups lacking effective rights and those who may be denying rights to others

(Conway et al., 2002). It maintains that a key objective of such an approach is to give both individuals and groups political, social, and economic power so that they are better positioned to take care of their own (rights-related) interests, and also proactively engage with the duty bearers, who are mostly the government agencies to better respond to the claims. This approach will however need to be coupled with additional activities like research, advocacy, evaluation, public education, and organizational development to make an impact on the lives and livelihoods of the poor who are the right holders.

10.1.5 Graduation Approach

If the goal is to end poverty, then we need to focus on the poorest of the poor who often get left behind in mainstream livelihood programmes as they are assetless and powerless. There is a need to look at the ultra-poor as a separate category having a unique set of needs, different from the poor (Hashemi & De Montesquiou, 2011). The BRAC graduation approach fills this development vacuum by designing a step-by-step development ladder for the ultra-poor to escape poverty. It involves a sequenced and time-bound set of interventions comprising of targeting, consumption support, savings, financial inclusion, skill transfer, asset transfer, life skills coaching, and microfinance. This approach was evaluated in a multi-country trial in diverse contexts and showed encouraging results (Banerjee et al., 2015). The cost per household is to the tune of US$ 550 per participant and is high due to frequent face-to-face coaching and asset transfer costs (Schelzig & Rawal, 2020). As a result, the risk of elite capture and resulting errors of inclusion is also high. However, certain elements such as the ease in estimating the project cost upfront, relatively assured outcomes, and a short graduation window of 2–3 years make this approach appealing to policymakers and donor agencies alike. Future challenges include harmonizing the per participant costs, adapting the interventions to the local context, and ensuring sustainability of the graduation outcomes. Since 2015, this approach has become popular and is being increasingly scaled up by donor agencies and governments.

10.1.6 Resilience Approach

Rising levels of shocks and stresses driven by natural disasters, social upheavals, economic crisis, political turmoil, pandemics, and climate change threaten the sustainability of our developmental achievements, resulting in a growing urgency of embedding resilience in development. Resilience is visualized as an outcome of three capacities, namely adaptive capacity, absorptive capacity, and transformative capacity in the face of shocks and stresses, and it exists at multiple scales, namely individual, household, community, society, etc. (Jeans et al., 2016). Several donor agencies are investing in strengthening disaster resilience, disaster risk reduction, and climate change adaptation. Resilience should not be looked upon as a goal in itself or a normative concept. Hence, resilience needs to be conceptualized not as a goal, but as a capacity building exercise to strengthen the ability of the vulnerable to respond to shocks and stresses without compromising on human wellbeing or dignity. Rapidly rising crisis contexts continue to challenge the human development achievements made during the MDG and SDG period, and development will now need to refocus on embedding resilience so that human wellbeing is sustained.

10.2 A Comparative Analysis

While assessing the comprehensiveness of the six livelihood approaches across select parameters, we note each of these approaches brings something unique to our understanding of rural livelihoods (Table 10.2). The macro-micro linkage, vulnerability context, and asset pentagon are the strength of the DFID SLA. Governing the commons framework highlights the importance of common pool resources, the role of institutions, and local enforcement. The livelihood trajectory approach helps assess the past and present livelihood status thereby establishing a baseline and the trend line for developing the future livelihood strategy. The rights-based approach to livelihoods builds on the normative principles of participation, inclusion, equality, non-discrimination, and accountability and

Table 10.2 Comparing the various livelihood approaches

Approach	Rationale	Applicability	Strength	Weakness
DFID SLA	A household decides its 'livelihood strategy' by combining its assets and taking into account its vulnerability context and macro policy environment	Universal applicability at the household level	Provides a conceptual framework for understanding the complexity of livelihoods	It is criticized as being too micro and too household focussed thereby limiting its ability to inform macro policy analysis. Due to its complexity, it is often expert-driven
Commons and livelihoods	Commons such as forests, pastures, fisheries, irrigation, and others when governed well can play a vital role in sustaining the livelihoods of the poor	In scenarios where the livelihoods of the poor are dependent on commons	Provides a framework for polycentric governance of the commons	While there is a lot of optimism in community-based governance of the commons, practical implementation has not always been successful
Livelihood trajectory	Livelihoods are inherently dynamic as households respond to shocks, stresses, and opportunities	In scenarios where livelihood trends need to be assessed to inform planning, monitoring, and impact assessment	Advocate that ending poverty will require two sets of interventions—one to accelerate ascent and the other to block descent	Livelihood trajectory studies are long term and retrospective
Rights-based approach	The root causes of generation and perpetuation of poverty and vulnerability are the unequal sharing of power and influence	Where there is structural discrimination, inequity, and marginalization of the local community	Can provide insights into the distribution of power, identify groups that lack effective rights, and those who may be denying rights to others	For marginalized groups, the provision of legal rights is not enough to ensure that they are respected by elites or enforced by the state
Graduation approach	Ultra-poor need to be looked upon as a separate category and treated differently from the poor	When the focus in on lifting the ultra-poor out of poverty in a time-bound manner	Provides a step-by-step sequential approach to lift the ultra-poor out of poverty in a time-bound manner	Affordability, errors of inclusion, and sustaining the livelihood gains
Resilience framework	Capacity building is needed for livelihoods to withstand shocks and stresses that are environmental, social, political, or economic	When in the face of sudden shocks and long-term stresses, human wellbeing needs to be sustained	Building adaptive, absorptive, and transformative capacity to sustain livelihoods in the face of risks and disasters	Need to be careful not to treat resilience as a normative concept, as it can be either good or bad just like governance

aims to provide individuals and institutions with social, economic, and political power so that they can better access their developmental rights. The graduation approach is strong on narrow targeting, market analysis, integrated livelihood strategy, and life skills coaching. The resilience framework focusses on the vulnerability context, macro context, building local capacity, and livelihood outcomes.

10.3 Creative Integration of the Approaches

In this section, we explore the possibility of integrating the livelihood approaches by building on their strengths to enhance our understanding of rural livelihoods. We discuss two integration methods, namely the convergent sequential where two or more livelihood approaches are applied in series one after the other, and the convergent parallel where the approaches are fused in parallel to provide greater clarity on understanding livelihoods.

10.3.1 Convergent Sequential

The vulnerability framework (exposure and sensitivity), the DFID SLA asset pentagon, and the resilience framework were integrated by Kaur et al. (2017). They look at resilience as an outcome instead of a capacity or ability. They assess how a national employment programme with its four main interventions of provisioning wages, developing infrastructure, decentralizing governance, and building capacity can assist households in strengthening their asset base to better respond to climate change risks and opportunities (Fig. 10.1). They found that 94% of the surveyed households had been able to absorb, adapt, or transform to address the impacts of climate change due to the programme interventions with only 4% reporting a decline in wellbeing.

The livelihood trajectory approach can be used to assess the pathways out of and into poverty. This understanding of how poor households escaped poverty in the past can inform the graduation approach in planning the future livelihood trajectories for the ultra-poor. Also, the graduation approach can be evaluated using the livelihood trajectory approach, to assess the sustainability of the graduation pathway. When prefixed to this approach, the livelihood trajectory approach can generate evidence on the livelihood pathways that are already functional and supporting the poor escape poverty. When suffixed, it informs whether the graduation pathway is sustainable or not. Also, the livelihood trajectory approach when prefixed to the resilience framework can provide insights on how households responded to shocks and stresses in the past, thereby providing guidance for the future.

The Household Livelihood Resilience Approach (HLRA) for measuring household-level resilience by quantifying the DFID SLA asset pentagon was developed by Quandt (2018). A set of indicators were used to quantify each of the five capitals (Table 10.3). These five capitals were then aggregated to construct the composite asset index which was used as a proxy for livelihood resilience. The DFID SLA and the rights-based approach can be combined synergistically. While the DFID SLA can identify the determinants of poverty at the micro-level, the rights-based approach can then at the macro-level engage in policy advocacy to formulate policies that provide a legal channel for claiming rights (Foresti et al., 2007).

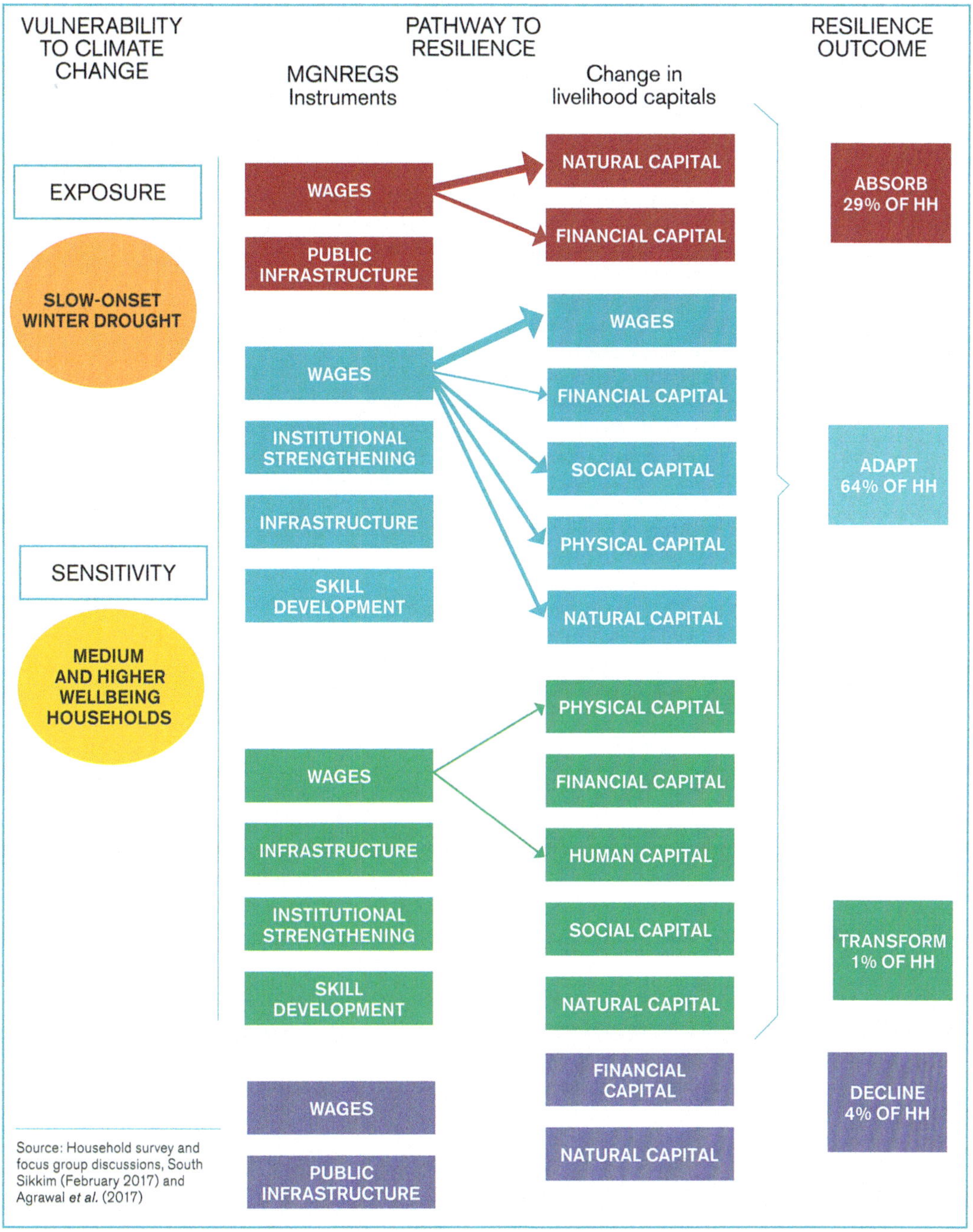

Fig. 10.1 Integrating vulnerability, asset pentagon, and resilience frameworks to assess the contribution of MGNREGS to household resilience. (Source: Kaur et al., 2017)

Table 10.3 List of livelihood resilience indicators at the household level (Source: Quandt, 2018)

Financial capital	• Salaried job (yes or no) • Access to a bank account (yes or no) • Remittances (yes or no) • Household belongings (# of belongings) • Livestock (# of livestock) • Size of farmland (# of acres) • Ownership of farm equipment (own, rent, borrow pieces of equipment)
Human capital	• Labour availability (number of household members between 18 and 55) • Education (level of education of respondent) • General health of family (scale of poor to good) • Health problems impact on ability to practice livelihoods (Scale of no to very much)
Social capital	• Family living nearby (yes, how close) • Political influence or power (scale of none to a lot) • Participation in groups (# of groups) • Participation in agriculture or tree planting group (yes or no) • Strength of relationship with neighbours (# of activities done with neighbours)
Physical capital	• Normal and rainy season road conditions (scale of good to bad) • Presence of facilities (schools, hospitals, etc.) near home (yes or no) • Access to irrigation schemes (yes or no) • Ownership of farming equipment (own, rent, borrow pieces of equipment)
Natural capital	• Size of farmland (# of acres) • Own farmland (yes or no) • Diversity of farm crops (# of different crops planted) • Livestock (# of livestock) • Soil erosion (rank of severity of soil erosion on farm)

10.3.2 Convergent Parallel

The livelihood strategy component in the DFID SLA can benefit when integrated with the graduation approach to help design a sequential set of interventions especially in the case of the ultra-poor. Both the approaches can benefit from this fusion, as the DFID SLA gains from a step-by-step articulation of the livelihood strategy, while the graduation approach gains from an added understanding of the vulnerability context, asset pentagon, and the macro-micro development context while planning the graduation. Integrating the collective governance of the commons approach with the DFID SLA, especially when assessing the asset pentagon and the institutional aspects will also add a lot of value, as the poor depend more on natural resources. Access to these CPRs needs to be acknowledged and accounted for as the general tendency is to include only privately owned assets while constructing the asset pentagon.

The rights-based approach can be useful in unpacking the policies, institutions, and processes (PIP) element of the DFID SLA by dovetailing the power analysis. While preparing a developmental plan and identifying interventions, it is vital to acknowledge and embed aspects of structural inequality, discrimination, inequality, and power relations that hinder the poor from improving their wellbeing (Foresti et al., 2007). The rights-based approach can also influence the way livelihood outcomes are monitored, from looking at only direct impacts, to also analysing whether it has led to empowering right holders in claiming their rights, strengthened the capacity of duty bearers to meet their obligations, and the change in the policy, legal, and institutional structures that enable the same (Foresti et al., 2007).

10.4 The 3PT Development Framework

What are the components that the bouquet of livelihood programmes in the rural development sector need to provision to comprehensively address the diverse requirements of a heterogeneous community? Devereux and Sabates-Wheeler (2004) developed the 3PT conceptual typology that separates the developmental interventions into four different categories—protection, prevention, promotion, and transformation (3PT) based on their objectives and the types of vulnerabilities they are trying to address (Summarized in Box 10.1).

Box 10.1 The 3PT Framework of Development

Protective Measures

Protective or provisioning measures are narrowly targeted safety net measures aiming to provide relief to marginalized individuals or groups such as children, orphans, elderly, or disabled people who are unable to work and earn their livelihood. This equates most closely to mainstream 'social welfare' programmes such as pension schemes, disability benefits, single-parent allowances, and 'social pensions' for the elderly poor. These programmes are financed publicly, with donor support, or through NGO projects. Other protective measures can be classified as social services. These would be for the poor and groups needing special care, including orphanages and reception centres for abandoned children, old-age homes, feeding camps and provision of services for refugees, internally displaced persons, and the waiver of health and education charges to extend access to basic services to the very poor. In the Disaster Risk Reduction context protective measures also include emergency feeding programmes, support for reconstruction, and restocking assets.

Preventive Measures

Preventive measures are defined as social or disaster linked policies and other safety net interventions that directly seek to reduce the vulnerability of individuals or groups to specific shocks and hazards through, for instance, unemployment schemes, insurance, or food and cash transfers. These are similar to 'social safety nets' and seek to avert deprivation, and reduce vulnerability. They include social insurance for 'economically vulnerable groups'—people who have fallen or might fall into poverty and may need support to manage livelihood shocks. These include formalized social insurance programmes of pensions, health insurance, maternity, and unemployment benefits, often with tripartite financing between employers, employees, and the state. They also include informal mechanisms, such as savings clubs and funeral societies. Strategies of risk diversification such as crop or income diversification are also considered preventive measures. More recently new forms of preventive measures concerning climate change adaptation and disaster risk reduction are emerging such as crop and weather insurance and health insurance to protect health and livelihood assets such as livestock.

Promotive Measures

Promotive measures aim to enhance real incomes and capabilities, and promote improved opportunities and livelihoods, through a range of livelihood-enhancing programmes targeted at households and individuals. They comprise public works (such as roads), infrastructure, access to credit, asset transfers and livelihood diversification support programmes, micro-credit, and others. These also include conditional cash transfers which incentivize investments in human capital by promoting demand for education and health and help address gender inequalities.

Transformative Measures

Transformative measures seek to address concerns of social equity and exclusion, such as collective action for workers' rights, or upholding human rights for minority ethnic groups.

(continued)

Box 10.1 (continued)

Transformative interventions include changes to the regulatory framework to protect 'socially vulnerable groups' (e.g. people with disabilities, or victims of domestic violence) against discrimination and abuse, as well as sensitization campaigns to transform public attitudes and behaviour and enhance social equity.

Source: Devereux and Sabates-Wheeler *(2004),* World Bank *(2012), Béné et al. (2012)*

This publication was originally published by the Institute of Development Studies (IDS). You can access it at: Transformative social protection (ids.ac.uk)

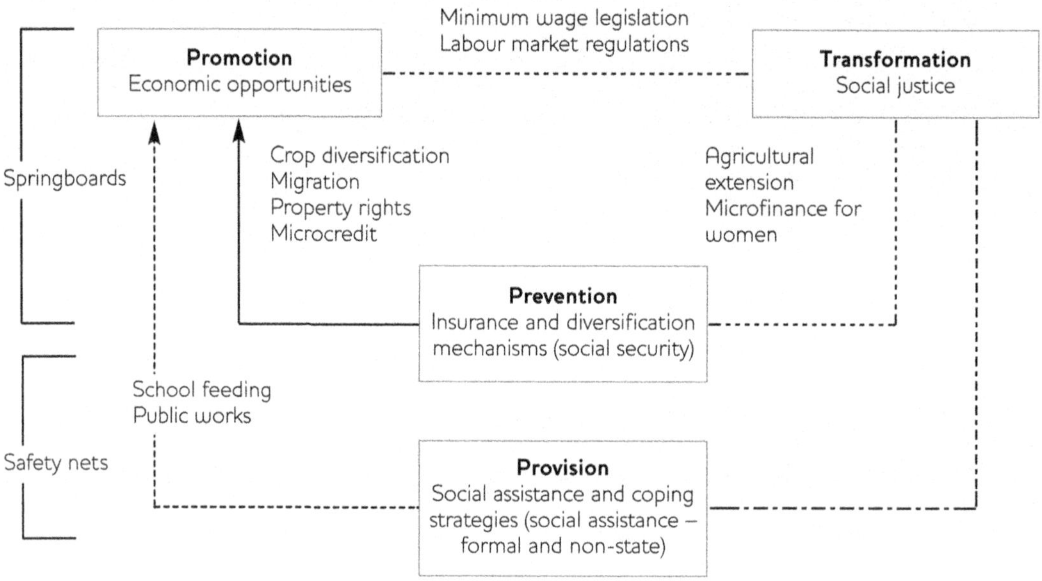

Fig. 10.2 The 3PT conceptual framework for social protection provides a useful basis for integrating the various livelihood approaches (Source: Sabates-Wheeler & Devereux, 2007). This publication was originally published by the Institute of Development Studies (IDS). You can access it at: Social Protection for Transformation | IDS Bulletin

The 3PT framework can be visualized as an overarching agenda to end poverty (Fig. 10.2) and the various livelihood approaches can be assessed to ascertain the extent to which they cover these four components using this lens (Béné et al., 2012). The livelihood approaches in their design emphasize different aspects of the 3PT framework (Table 10.4). The DFID SLA focusses on the agency of the poor to reduce their vulnerabilities, strengthen their asset mix, and progress towards sustainable livelihoods. The common pool resources approach highlights the role of commons in enabling a safety net, promoting livelihoods, and also governing them through common property regimes. The livelihood trajectory approach emphasizes the need to prevent descent into and promote escape out of poverty. The rights-based approach focusses on the legal rights to transform livelihood security and social justice. The graduation approach focusses on safety nets and springboards in a sequenced manner to assist the ultra-poor escape poverty. The resilience framework emphasizes building capacities so that households can cope, adapt, and transform their livelihoods in the face of shocks and stresses. This analysis using the 3PT lens shows that the emphasis of the livelihood approaches is mainly on prevention and promotion, while the protective (or provisioning) and transformative dimensions have received less attention. Also, integration of the livelihood approaches can help provision a complete livelihood package for the poor.

Table 10.4 Analysing the various livelihood approaches using the 3PT lens

RLA	Protection	Prevention	Promotion	Transformation
DFID SLA		✓	✓	
Commons		✓	✓	✓
Trajectory		✓	✓	
Rights-based	✓	✓	✓	✓
Graduation		✓	✓	
Resilience	✓	✓	✓	✓

10.5 The Takeaways

Each of the six livelihood approaches discussed in this chapter adds a unique dimension to our understanding of rural livelihoods. A deeper understanding of their niche area, functionality, and ease of application can guide in creating the best fit in a given situation. We should not frame these six livelihood approaches as being mutually exclusive and distinct from each other, but instead, conceptualize them as a family of livelihood approaches. A fusion of these approaches can be explored in a manner such that they complement each other. These diverse livelihood approaches while occupying their niche areas can be integrated, thereby providing a more nuanced and complete understanding of rural livelihoods in research, policy, and practice.

Exercises

1. Describe how the poverty alleviation approach of India's National Rural Livelihoods Mission (https://aajeevika.gov.in/) differs from that of the BRAC graduation approach. Compare and contrast these programs using the 3PT framework to assess their ability to deliver a comprehensive package to the poor.
2. Study the research paper on 'Poverty pathways in Andhra Pradesh' by Krishna (2006) and discuss whether the poverty dynamics approach provides an in-depth analysis of rural livelihoods and its implications for development planning? How would you creatively supplement it with the other livelihood approaches to further enrich the picture? https://www.findevgateway.org/sites/default/files/publications/files/mfg-en-pathways-out-of-and-into-poverty-in-36-villages-of-andhra-pradesh-india-2005.pdf
3. The Mahatma Gandhi National Rural Employment Guarantee Act (MGNREGA) of India owes its genesis to the rights-based approach and was created by transforming the erstwhile public works programme to a legal guarantee of the right to employment. While assessing the efficacy of a national employment guarantee programme in addressing climate change impacts, Kaur et al. (2017) integrated the three frameworks, namely DFID SLA, vulnerability, and resilience. What value did this sequential integration add to the overall conceptualization of the study? https://archive.nyu.edu/jspui/bitstream/2451/44184/2/Building%20resilience.pdf
4. Discuss whether the BRAC graduation approach provides a sure shot strategy to lift the poorest out of poverty? What are the gaps you perceive in this approach, and how would you creatively use the other livelihood approaches to further strengthen this approach?
5. Study the design of the Bihar Rural Livelihoods Project approach (JEEViKA, 2011) and how it addresses social exclusion to improve livelihoods of the rural poor in Bihar, India. Discuss how the other livelihood approaches can be integrated with this project to strengthen its outcomes. http://documents1.worldbank.org/curated/en/298391515516507115/122290272_20180012032255/additional/122548-WP-P090764-PUBLIC-India-BRLP-Booklet-p.pdf
6. Visit the website https://www.ifad.org/en/web/latest/stories- to select one project story and re-write the story by embedding various livelihood approaches in it. The only thing you need to

ensure is that no two students should select the same story. Indicate the IFAD URL of the story in your assignment.

7. An impact evaluation of a public microfinance programme was carried out by integrating the BRAC graduation approach in the evaluation methodology (Tambe et al., 2017). Discuss how the graduation approach was embedded in the evaluation design and what value it added in the overall design.
8. The BRAC graduation approach along with participatory rural appraisal (PRA) tools was integrated into the planning phase of a public wage employment programme to enable it to create assets in the lands of the poor (Tambe et al., 2019). Discuss how the graduation approach was embedded in the programme and what value it added to the programme outcomes.
9. Participatory Wealth Ranking (PWR) in a village reveals that there are six categories of households, namely destitute, extreme poor, moderate poor, vulnerable non-poor, non-poor, and the wealthy as indicated in the figure below (sourced from De Montesquiou et al., 2014, 14). The interventions available in your project are consumption support, savings, skill transfer, asset transfer, life skills coaching, and microfinance. Illustrate which of these interventions you will use and in what sequence to graduate the households in each of these categories in an efficient and time-bound manner. Also, discuss how you would provision the components of the 3PT development framework differentially to the various categories of the households in an efficient manner using a public funded programme.

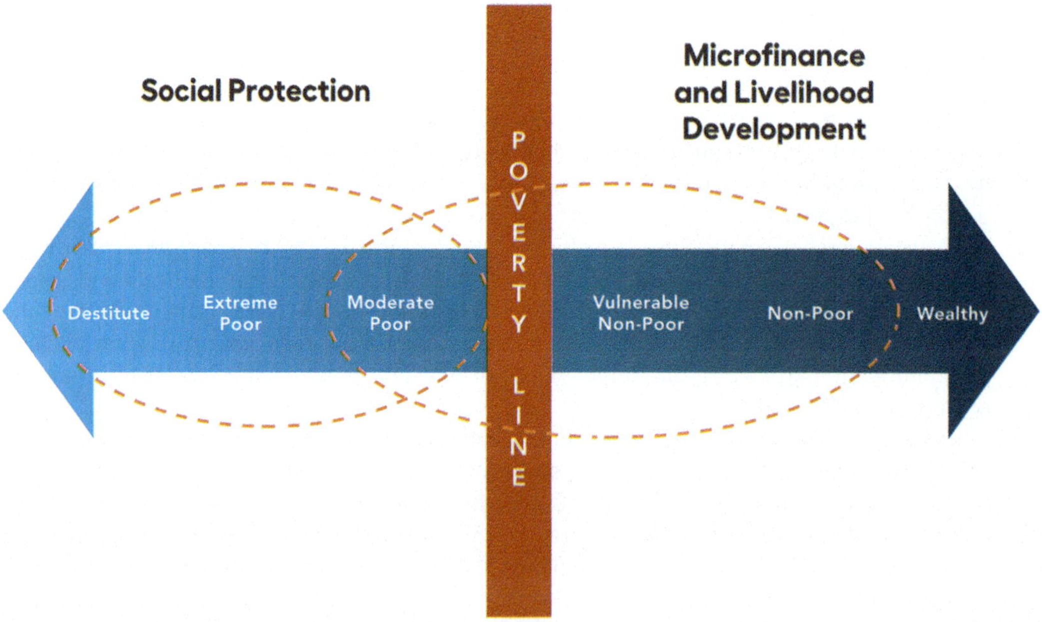

https://www.cgap.org/sites/default/files/graduation_guide_final.pdf

10. Béné et al. (2012) integrate the 3D resilience framework (absorb, adapt, transform) with the Protection-Prevention-Promotion-Transformation (3PT) framework to develop the 3PT-3D framework. They apply this framework to assess the BRAC Targeting the Ultra-Poor (TUP) graduation programme in Bangladesh. Critically discuss the findings of this analysis and its implications for designing development programs. https://onlinelibrary.wiley.com/doi/pdfdirect/10.1111/j.2040-0209.2012.00405.x

References

Banerjee, A., Duflo, E., Goldberg, N., Karlan, D., Osei, R., Parienté, W., et al. (2015). A multifaceted program causes lasting progress for the very poor: Evidence from six countries. *Science, 348*(6236).

Béné, C., Wood, R. G., Newsham, A., & Davies, M. (2012). Resilience: new utopia or new tyranny? Reflection about the potentials and limits of the concept of resilience in relation to vulnerability reduction programmes. *IDS Working Papers, 2012*(405), 1–61.

Conway, T., Moser, C., Norton, A., & Farrington, J. (2002). *Rights and livelihoods approaches: Exploring policy dimensions*. Natural Resource Perspectives, The Overseas Development Institute.

De Montesquiou, A., Sheldon, T., Degiovanni, F., & Hashemi, S. (2014). *From extreme poverty to sustainable livelihoods: A technical guide to the graduation approach*. CGAP and Ford Foundation. Retrieved August 16, 2020 at 9:00, from https://www.cgap.org/sites/default/files/graduation_guide_final.pdf.

de Silva, T. E. (2013). *Principles in international development: Sustainable livelihoods and human rights-based approaches*. Division of Economic and Labor Research Office of Trade and Labor Affairs U.S. Department of Labor, USA.

Devereux, S., & Sabates-Wheeler, R. (2004). *Transformative SOCIAL PROTECTION*. IDS Working Paper 232, Brighton: Institute of Development Studies. Transformative social protection (ids.ac.uk).

DFID. (1999). *Sustainable livelihoods guidance sheets, numbers 1-10*. Department for International Development, London, UK.

Dorward, A., Anderson, S., Bernal, Y. N., Vera, E. S., Rushton, J., Pattison, J., & Paz, R. (2009). Hanging in, stepping up and stepping out: Livelihood aspirations and strategies of the poor. *Development in Practice, 19*(2), 240–247.

Foresti, M., Ludi E., & Griffiths, R. (2007). *Human rights and livelihood approaches for poverty reduction*. ODI Paper prepared for the SDC Poverty-Wellbeing Platform. Retrieved June 5, 2021 at 10:45 am, from https://cdn.odi.org/media/documents/HR_livelihood_approaches.pdf.

Hashemi, S. M., & De Montesquiou, A. (2011). Reaching the poorest: Lessons from the graduation model. *CGAP, Focus Note, 69*(1), 1–15.

Jeans, H., Thomas, S., & Castillo, G. (2016). *The future is a choice: The Oxfam framework and guidance for resilient development*. Oxfam International.

JEEViKA. (2011). *Addressing social exclusion to improve livelihoods of rural poor in Bihar*. JEEViKA Learning Note Series, No. 2, World Bank. Retrieved from http://documents1.worldbank.org/curated/en/298391515516507115/122290272_20180012032255/additional/122548-WP-P090764-PUBLIC-India-BRLP-Booklet-p.pdf.

Jones, L., & Tanner, T. (2015). *Measuring subjective resilience: Using peoples' perceptions to quantify household resilience*. Overseas Development Institute, Working Paper 423. https://doi.org/10.13140/RG.2.1.2438.1289.

Kaur, N., Steinbach, D., Agrawal, A., Manuel, C., Saigal, S., Panjiyar, A., Shakya, C., & Norton, A. (2017). *Building resilience to climate change: MGNREGS and climate-induced droughts in Sikkim*. IIED Issue Paper, IIED.

Kothari, U., & Hulme, D. (2004). *Narratives, stories and tales: Understanding poverty dynamics through life histories*. GPRG-WPS-011. ESRC Global Policy Research Group, Manchester, 41pp.

Krishna, A. (2006). Pathways out of and into poverty in 36 villages of Andhra Pradesh, India. *World Development, 34*(2), 271–288.

Offenheiser, R. C., & Holcombe, S. H. (2003). Challenges and opportunities in implementing a rights-based approach to development: An Oxfam America perspective. *Nonprofit and Voluntary Sector Quarterly, 32*(2), 268–301.

Ostrom, E. (1990). *Governing the commons: The evolution of institutions for collective action*. Cambridge University Press.

Quandt, A. (2018). Measuring livelihood resilience: The household livelihood resilience approach (HLRA). *World Development, 107*, 253–263.

Sabates-Wheeler, R., & Devereux, S. (2007). Social protection for transformation. *IDS Bulletin, 38*(3), 23–28. https://doi.org/10.1111/j.1759-5436.2007.tb00368.x

Schelzig, K., & Rawal, P. (2020). *Testing the GRADUATION APPROACH in the Philippines*. Asian Development Bank, Briefs No. 120, Manila, Philippines.

Scoones, I. (1998). *Sustainable rural livelihoods: A framework for analysis*. Working Paper 72, Institute for Development Studies, Brighton, UK.

Tambe, S., Bhutia, N. T., Pradhan, S., & Basi, J. (2019). Coupling a ladder to the safety net: Reinventing MGNREGA for asset creation. *Development in Practice, 29*(4), 514–524. https://doi.org/10.1080/09614524.2019.1567687

Tambe, S., Patnaik, S., Bisaria, J., Pandey, A., Sarma, U. K., & Ananda Vadivelu, G. (2017). How does government microfinance impact the rural poor? Evidence from Madhya Pradesh (December 30, 2017). *Economic and Political Weekly, LII*(52), 81–87.

World Bank. (2012). *Resilience, equity and opportunity, The World Bank's social protection and labour strategy 2012–2022*. Washington, DC. Retrieved April 21, 2021 at 21:00, from http://documents1.worldbank.org/curated/en/443791468157506768/pdf/732350BR0CODE200doc0version0REVISED.pdf.

Wunder, S. (2011). Poverty alleviation and tropical forests—What scope for synergies? *World Development, 29*(11), 1817–1833.

Part III
Strengthening the Knowledge Action Pathway

The practice of rural livelihoods in both research and development is dominated by the use of generalist approaches. Can the six livelihood approaches that we have studied, namely the sustainable livelihoods approach, collective governance of the common pool resources, livelihood trajectories, rights-based approach, graduation approach, and the resilience framework better inform the science, policy, and practice of rural livelihoods? What are the situations that researchers and development professionals face and how can the application of rural livelihood approaches add value? We engage with this question in two parts. Firstly, in Chap. **11**, we discuss how to embed the livelihood approaches directly in practice, and further in Chap. **12**, we explore the possibility of mainstreaming these approaches in existing practice that is dominated by generalist project management approaches. In the last chapter, we discuss the past, present, and future of rural livelihoods in theory and practice. We opine that the family of six livelihood approaches need to be visualized as a family of approaches and their nuanced understanding and suitable application can go a long way in informing rural livelihoods, and also contribute to the SDG goal of ending poverty and hunger by **2030** and sustaining it in future.

Chapter 11
Embedding Livelihood Approaches in Science, Policy, and Practice

Abstract In this chapter, we engage with the needs of researchers, policymakers, and practitioners. Livelihood research is currently dominated by mixed methods followed by statistical analysis. We provide guidance on 'how to' integrate livelihood approaches in academics. We also explore the possibilities of embedding the livelihood approaches in project planning, monitoring, and evaluation. During project planning, livelihood trajectories can help discern the pathways into and out of poverty, thereby informing project strategy. The governing the commons approach can guide the sustainable management of the common pool resources. The rights-based approach can help in unearthing the root causes of poverty that hamper communities from exercising their rights and freedoms. The graduation approach provides a set of sequenced developmental inputs to lift the ultra-poor out of poverty. Also, the sustainability of the project outcomes can be strengthened using the resilience framework. In project monitoring and evaluation, rural livelihood approaches can guide in assessing the baseline and endline, constructing the key performance indicators, and ascertaining the attribution. The conceptualization of a unified livelihood framework is also explored. We opine that the family of rural livelihood approaches have a lot to offer in achieving the SDG goals of ending poverty and hunger.

Keywords Rural worlds · Livelihood analysis · Prepare proposals · Livelihood plan · Evidence-based policy · Project monitoring · Evaluation · Comprehensive element · Specialized element · Unified approach

11.1 Understanding Poverty and Livelihoods

The science, policy, and practice of poverty and livelihoods can gain a lot from the knowledge and experience gained over the last few decades. A better understanding of 'how' poverty is locally perceived, 'who' is poor, 'why' they are poor, and 'what' needs to be done is at the heart of designing effective poverty alleviation programmes. Matin and Hulme (2003) point out that over the years there have been three key advances in the ideas that have impacted poverty reduction thinking. First, is the recognition that the poor are not a homogenous group, second, it is now recognized that sustainable poverty reduction requires both a protection and a promotional component, and thirdly, that prescriptive or top-down approaches are likely to fail. Scoones (2015) highlights six key questions that one should be asking in a livelihood analysis. Who owns what or has access to what? who does what? who gets what? what do they do with it? How do social classes and groups interact with each other? How do changes in politics get shaped by dynamic ecologies and vice versa? Livelihood studies and programmes need to acknowledge inter and intra-village heterogeneity, the diverse assets, and livelihood strategy of the different household groups, and predict how livelihood diversification of one group might impact the others. OECD (2007) developed the typology of five rural worlds (5RW) that segregates the rural households into five stylized types of households and is a useful contribution on this front (Box 11.1).

S. Tambe, *Teaching and Learning Rural Livelihoods*, Sustainable Development Goals Series,
https://doi.org/10.1007/978-3-030-90491-3_11

Box 11.1 The Typology of the Five Rural Worlds (5RW)
The current reality in rural areas is defined by a highly diverse range of stakeholders involved in agriculture with considerable variation in their assets, access to markets, and institutional systems. A typology of five rural worlds (5RW) can guide policymakers in understanding this diversity and respond with appropriate pro-poor policies and is presented below with some modifications (OECD, 2007):

- *Rural World 1:* Large-scale commercial agricultural households and enterprises with an influential voice in national policies, political affairs, and ties to buyer-driven value chains. These enterprises also employ members of the other worlds.
- *Rural World 2:* Traditional agricultural households and enterprises, not internationally competitive but with sizeable landholdings devoted to both commercial and subsistence production.
- *Rural World 3:* Agricultural households and micro-enterprises, including those with limited assets (land, credit), living in fragile ecosystems, with small units producing food for consumption and sale on local markets, and are dependent on off-farm employment for a significant part of their incomes.
- *Rural World 4:* Landless rural households and micro-enterprises, whose labour and community ties are their major assets and sharecropping and migration are main livelihood options. This World relies on Rural Worlds 1 and 2 for employment and income-generating opportunities.
- *Rural World 5:* Chronically poor rural households, many of which are no longer economically active, and have sold off or have lost their assets during periods of crisis. Social exclusion and gender inequalities are rife and safety nets for this group are essential.

Source: OECD (2007), Castillo (2008)

11.2 Integrating RLAs in Research

Researchers often need to take up livelihood analysis, measure the contribution of particular resources to livelihoods, uncover the heterogeneity of human wellbeing, unearth the dynamics of poverty, analyse livelihood patterns, discern the drivers, etc. Livelihood research primarily relies on field observation, household surveys, and participatory group exercises and can benefit from the insights that livelihood approaches offer. Livelihood and poverty studies can gain by riding on the RLAs as the framework or lens with which to analyse the livelihood patterns. We discuss a few examples of how RLAs can be integrated with livelihood research and the value they can add. Amongst the RLAs, the DFID sustainable livelihood approach is the most comprehensive and has been comparatively used much more in research. We analyse a few studies that have used these approaches and highlight the best practices, pitfalls, and learnings.

Masud et al. (2016) carried out a livelihood analysis of the marine national park communities in Malaysia. They collected primary data using household interviews and focus group discussions to construct the DFID SLA and provide useful insights on the asset pentagon, livelihood strategy, and the vulnerability context (Box 11.2). We opine that this study could have further benefitted by incorporating intra-village heterogeneity using the participatory wellbeing ranking (PWR) tool to provide a more inclusive picture of livelihoods by assessing it separately for the different household groups.

The other elements of the livelihood trajectory approach such as the transition matrix and life stories (detailed in Sect. 6.3.2, Table 6.3, and Box 6.2 of Chap. 6) would have benefitted the study by assessing the poverty dynamics. Several other livelihood research studies have benefitted from applying the DFID SLA. Ellis and Freeman (2004) conducted a livelihood assessment study in four east African countries. They started with a participatory wellbeing ranking to develop a criterion to segregate households into well off, middle class, and poor based on their land ownership, livestock holding, and vulnerability status. They then developed a comparative asset pentagon for these categories by quantifying select assets. The study incorporates intra-village heterogeneity and aggregates the primary data to construct the asset pentagon for the various household categories to discern the factors that enable or impede poverty reduction.

Box 11.2 Assessing Livelihoods in the Marine Areas of Malaysia using DFID SLA
Masud et al. (2016) took up a research study to assess the livelihoods of communities located in the marine national park areas of peninsular Malaysia. As the study area was large they used sampling techniques to identify the villages and then used systematic random sampling to select every third household in these identified villages. A field study was carried out using mixed methods comprising of a household questionnaire and focus group discussions to assess the livelihoods of communities residing in the marine national park areas in peninsular Malaysia. This data was then used to populate the DFID SLA, with the asset pentagon and livelihood strategy constructed using the household survey, and the vulnerability context using both the household survey and the focus group discussions. Several livelihood studies use mixed methods but do not use the RLA lens to conceptualize their study. This study is a good example of integrating mixed methods with a RLA to provide deeper insights. This study could have further benefited by integrating the livelihood trajectory approach. Acknowledging village heterogeneity, it could have started with the participatory wellbeing (PWR) ranking to segregate the households in the identified villages into three or four categories (very poor, poor, middle class, well off) followed by stratified random sampling within each of these categories, instead of the systematic random sampling. We opine that this study would have been more nuanced and inclusive this way, as the findings would have been generated separately for each of these wellbeing categories. After the PWR exercise, the livelihood trajectory approach could have helped generate the transition matrix to illustrate the poverty pathways followed by the qualitative household interviews to ascertain the enabling and impeding factors.

Source: Adapted from Masud et al. (2016)

11.3 RLAs Informing Policy

The rural livelihood approaches can help identify backward or under-developed regions where geographical targeting of policy interventions would be needed. Donohue and Biggs (2015) use the DFID SLA to assess the livelihood status at the sub-national level. They do so by aggregating the 23 socio-economic indicators to construct the asset pentagon (natural, physical, financial, human, and social) across the 13 sub-regions to identify geographical backwardness needing policy priority. The common pool resources (CPR) approach by building on empirical studies around the globe highlights that community-based governance of the commons need not always result in 'tragedy of the commons' leading us to advocate for government or private control. A polycentric governance model can work where local communities govern the commons sustainably with the state providing an enabling external environment (funds, infrastructure, markets, etc.) without interfering in local level management or decision making. The livelihood trajectory approach provides guidance to poverty policy by advocating that two sets of policies will be needed to end poverty, one to accelerate the ascent out of poverty and the other to block the descent into poverty (Box 11.3). It also goes on to add that other than identifying who is poor, the reasons behind people escaping and falling back into poverty need to be unearthed to inform policymaking. The graduation approach advocates policy to treat the ultra-poor as a separate category different from the poor, with specific needs and needing a separate set of policy interventions. This approach is especially relevant in our quest to end poverty and hunger by 2030. We provide a useful example from the Sikkim state of India of how the asset-based approach of DFID SLA and graduation was embedded in an employment programme to make it more effective and pro-poor (Box 11.4). The rights-based approach challenges policymakers to delve into the root causes of poverty which are often structural and the need to tackle them with policy and legal tools by empowering right-holders to claim their rights and enabling duty bearers to fulfil their obligations. The resilience approach can guide policy on how to safeguard livelihoods from shocks and stresses to ensure that the developmental gains are sustained.

Box 11.3 Poverty Pathways Informing Developmental Policy

Krishna (2007) points out that two sets of policies are needed to end poverty, one set to prevent people from falling into poverty, and another set to promote escapes out of poverty. Both sets of policies are needed simultaneously as the first set will ensure that the non-poor do not become poor while the second set will assist those who are presently poor. Also, policymakers need to be mindful that there are separate reasons for escape and descent into poverty, which need to be tackled differently but simultaneously. A series of studies revealed that 90% of the households that have come out of poverty are due to improved yields from agriculture, jobs in the informal sector, and full-time, secure jobs in the private and public sectors. While ill health and high healthcare costs are the main reasons for a household's descent into poverty, social and customary expenses on marriages and funerals constitute another set of factors often associated with descent.

Box 11.4 Asset-Based Graduation Informing India's Employment Program

The Mahatma Gandhi National Rural Employment Guarantee Act (MGNREGA) was launched in 2005 and soon grew into the largest anti-poverty and public works programme in the world. It guaranteed unskilled wage employment to every rural adult demanding it. The public opinion on this programme was polarized. While some perceived it as wasteful public expenditure, others considered it as a lifeline for the poor. In the Sikkim state, located in the north-eastern region of India, the officials heading this programme observed that while creating public assets (such as village footpaths, playgrounds, drainage works, torrent control, etc.) resulted in wage employment for the poor households, however, it was not sufficient to lift them out of poverty. They wondered how to restructure this safety net so that it could also double up as a livelihood ladder without altering its entitlement-based framework. They decided to create household livelihood assets (HLAs) in the lands of the poor. These HLAs could take the form of cattle sheds, water storage tanks, land terracing, horticulture orchards, fodder development, and the like. The main hurdle this strategy faced was that the village-level elected representatives preferred public assets as it was easier for them to showcase their achievements. They were also apprehensive that targeting only the poor would negatively impact their popularity. These HLAs would also require a higher investment in procuring materials (cement, steel, stone, sand, etc.), while the

(continued)

Box 11.4 (continued)

programme capped the expenditure on the material component at 40%. Identifying the poor households and the most suitable HLAs for them would need a village-level planning process to be put in place.

To accelerate the creation of HLAs in the lands of the poor, a four-pronged strategy was conceptualized. A state-level, pro-poor policy was enforced that mandated that at least half of the village MGNREGA funds shall be invested in the lands of the poor. Convergence with line departments (such as agriculture and horticulture) was formalized to leverage the material component and technical support from them. Village-level planning was grounded using a participatory wealth ranking (PWR) tool to identify the poor and the strengths, weaknesses, opportunities, and threats (SWOT) analysis was done at the household level to identify the most suitable HLA. Finally, effective social audits were put in place to detect leakage and take corrective action.

It took 4 years of effort to ground this strategy (2012–2016), and the shelf of projects shiftED from public construction works to livelihood assets for the poor. The investment in HLAs increased five times from 10 to 52%. Animal husbandry which is the main livelihood for the landless and near landless households was prioritized for the first time. Works related to horticulture plantations (large cardamom and mandarin orange), fodder development, and land terracing were also scaled up. Also, a third of the state MGNREGA budget shifted to convergence mode, bringing in the much-needed investment in the material component and technical support from the line departments.

An independent impact evaluation found that the greatest multiplier effect was seen in the case of large cardamom plantations, followed closely by orange plantations and cowsheds. Also, the creation of these household livelihood assets contributed towards a self-reliant rural economy that appeared to move towards self-employment by lowering economic dependence on government programs.

Source: Tambe et al. (2019)

11.4 Applying RLAs in Planning

There are many unforeseen real-life challenges that development professionals face. Often, they need to prepare project proposals that range across diverse thematic areas such as poverty alleviation, promoting sustainable livelihoods, or more focussed thematic projects on sustainable farming, watershed development, village enterprise, value chain development, good governance, women-led collectives, climate change adaptation, and the like. We describe a step-by-step procedure on how to combine the rural livelihood approaches in project planning. While preparing a livelihood plan for a village, first the households need to be stratified using the participatory wellbeing ranking tool and the livelihood approaches applied separately for each of these categories. A unified livelihood framework can be conceptualized with the DFID SLA as the comprehensive element, and the other RLAs embedded in specific contexts as specialized elements to add value. To implement this, first a participatory assessment of the livelihood context comprising of the first three elements of the DFID SLA (namely the vulnerability context, asset pentagon, and the macro policy, institutions, and processes) is carried out in group exercise mode. Based on the livelihood context, the suitable specialized RLAs (commons, trajectory, graduation, rights-based, or resilience) are embedded to sharpen the livelihood strategy element of DFID SLA as detailed in Table 11.1. Where commons (forests, fishery, etc.) play an important role, one can draw on the Ostrom's governing the commons approach. Where past trends matter, the livelihood trajectory approach can help discern the poverty pathways. If the ultra-poor households

Table 11.1 Based on the project objectives and specific livelihood context, the specialized RLAs can be embedded to add value to the 'livelihood strategy' element of the DFID SLA, thereby developing a unified livelihood framework

Livelihood context	Suitable livelihood approach	Value it adds to livelihood strategy
Livelihoods dependent on common pool resources	Commons	Guidance on governing the commons sustainably
Need to build on the past and present livelihood trends	Trajectory	Discerning the pathways out of and into poverty over the last few decades
Ultra-poor getting left behind	Graduation	Prescribes a step-by-step, sequential approach to lift the ultra-poor out of poverty
Deep rooted exploitation and discrimination of the marginalized	Rights-based	Brings back politics in poverty discourses with a focus on structural inequalities
Crisis context with rising shocks and stresses	Resilience	Focus on building absorptive, adaptive, and transformative capacity to deal with adversity

Table 11.2 Unique elements that the RLAs bring to livelihoods thinking

Livelihood approach	Unique elements
DFID SLA	Vulnerability context, asset pentagon, macro policy, institutions, and processes
Livelihood trajectory	Assess the past trends, transition matrix, poverty pathways, life stories, the need to block descent, and accelerate the ascent
Common pool resources	Contribution of commons to livelihoods, common property regimes, design principles of common pool resources (Box 5.2)
Graduation	Life coach, market analysis, telescopic vision of the ultra-poor, basket of feasible livelihood options for the poor, asset transfer
Rights-based	Power relations, the root cause of poverty, structural inequalities, role of policy and law in addressing discrimination
Resilience	Shocks and stress, capacity to absorb, adapt, and transform in the face of change and disruption

are being left behind, one may embed the graduation approach. If discrimination and exclusion are the root cause, then the rights-based approach can help in empowering the stakeholders. Invulnerable sites are frequently affected by shocks and stresses, the resilience framework can guide livelihood strategy to build absorptive, adaptive, and transformative capacity. In this manner, based on the local livelihood context unearthed by the DFID SLA, the most suitable RLA can be identified and integrated to guide livelihood strategy. We opine that an integrated approach where we build on the strengths of the RLAs and embed them according to the context can give the best results. Each of these RLAs occupies a niche area and comprised of unique elements, and based on the context have the potential to add a distinctive value to the livelihood analysis as indicated in Table 11.2.

11.5 RLAs in Monitoring and Evaluation

Monitoring and evaluation help to assess the extent to which the project outcomes were met, how human wellbeing has changed, who has benefitted and who has not, how the asset profile has changed, and more importantly whether it was driven by the project intervention to confirm attribution. Frameworks that incorporate the change dimension such as the DFID SLA, livelihood trajectory, and resilience are especially suited for this purpose. DFID SLA is very relevant for the before and after analysis as well as the process of change. It is especially useful to theorize how households benefit from the interventions in terms of attainment of livelihood outcomes and how it results in an expansion of the asset pentagon. We demonstrate the use of DFID SLA in evaluation with the help of a few studies. Qian et al. (2017) analysed the impact of two different models of tourism, namely community-based tourism and the lease-to-operate tourism on livelihoods in two villages of rural China. They used mixed methods combining quantitative household surveys with key informant interviews and non-participant observation. Using this data, they quantified 15 indicators to construct the asset pentagon of these two villages both before and after the introduction of tourism. Though the study assumes the households as homogenous, it serves as a useful example of applying the DFID SLA in evaluation. We opine that stratification of the households using participatory wellbeing ranking would have enabled the study to ascertain the equity aspects as well that is the extent to which tourism development benefitted the poorest.

The livelihood trajectory approach can also be used to assess the progress and attribution. The control village is the counterfactual and has attributes similar to the treated village. The transition matrix in the treated and control village can be assessed and compared. This matrix will indicate the extent to which the very poor and poor households have progressed and whether their pathway can be attributed to the intervention. This approach is very useful in instances where project baselines are weak, as they can be recreated post facto in the treated and control villages based on a recall survey. The qualitative aspect is a narrative of the household's journey and can strengthen attribution, as one can ascertain the drivers of change while documenting the life stories.

11.6 Learnings and the Way Forward

The rural livelihood approaches can play an important role in strengthening livelihood research, livelihood policy, and in informing project planning, monitoring, and evaluation. Livelihood research often needs to take up livelihood assessment and hence the DFID SLA and the livelihood trajectory approach have an important role in academics. While in planning, as the livelihood strategy needs to be framed, we conceptualize a unified livelihood framework by looking at DFID SLA as the comprehensive element and the other RLAs as the specialized elements to be used based on the livelihood context. One can thus integrate the livelihood approaches in this manner and get the most out of the generalist and specialist livelihood approaches. In project monitoring and evaluation, as change has to be measured, the DFID SLA, livelihood trajectory, and resilience frameworks can add immense value. We opine that while the livelihood approaches have a lot to offer to the science, policy, and practice of rural livelihoods, they are not adequately used and we hope that this book will contribute towards changing the trend in the near future.

Exercises

1. A livelihood project was initiated in the year 2010, to raise household incomes and adopted a prospective evaluation approach. It gathered data on both the treated households and the comparison group from the beginning. In the year 2018, on completion of the project, the end-project evaluation came up with the following findings:

Household income levels	2010	2018
Treated household	Rs. 4000	Rs. 8000
Comparison household	Rs. 3000	Rs. 8000

- What is the change in the outcome indicator (household incomes)?
- What is the impact of the project?
- Can you provide a real-life instance where this scenario is possible?

2. Describe how the poverty alleviation approach of the National Rural Livelihoods Mission (DAY-NRLM) of India differs from that of the BRAC graduation approach. Cover the aspects of targeting, consumption support, savings, skill transfer, asset transfer, life skills coaching, microfinance, women collectives, institutions of the poor, etc.
3. Thanh et al. (2020) study the differential livelihood trajectories in coastal communities in Vietnam where transformations from artisanal fishing to aquaculture are now ubiquitous. They identify accumulating, fluctuating, and marginalizing as the three distinct livelihood trajectories available to the fishing and aquaculture-dependent households. While the transformation to aquaculture has benefitted the adopters, it has harmed the livelihoods of the artisanal fisher folks and the ecosystem. Based on the learnings of this study, frame a suitable evidence-based policy for fishery management for the coastal fishing communities. https://doi.org/10.1016/j.worlddev.2020.105219
4. Climate change related vulnerability is determined by exposure, adaptive capacity, and sensitivity. Tambe et al. (2011) construct the climate-related vulnerability index of the villages in the Sikkim Himalaya by using this measure. Examine the indicators used by them to construct this index, and the recommendations of the study. Based on this, prepare the policy recommendations to inform developmental planning in the state. https://www.currentscience.ac.in/Volumes/101/02/0165.pdf
5. Cash et al. (2003) provide a compelling account of why some science is translated into action while others are not. They demonstrate how in the CGIAR system at the International Maize and Wheat Improvement Centre (CIMMYT) attempts at the successful translation of new research findings on crop breeding were not adopted by farmers. Analyse the study using the credibility, relevance, and legitimacy (CRELE) framework and the importance of communication, translation, and mediation in strengthening the role that science and technology can play in sustainable development. https://doi.org/10.1073/pnas.1231332100
6. Tambe et al. (2019) share their experiences of bridging the science, policy, and practice interface in the context of the eastern Himalaya. They present three case studies dealing with pastoral livelihoods, promoting sustainable rural livelihoods, and co-designing a Himalayan spring revival initiative. Discuss how the credibility, relevance, and legitimacy (CRELE) of the knowledge produced was strengthened to translate the science into action. https://doi.org/10.1007/s10668-019-00456-8
7. Tambe et al. (2017) carry out an impact evaluation of a public microfinance programme—the National Rural Livelihood Mission (NRLM) in the state of Madhya Pradesh in India. Based on the recommendations of the study, what evidence-based policy improvements would you suggest in the design of the NRLM programme? The mission document of this programme is available at https://aajeevika.gov.in/sites/default/files/nrlp_repository/nrlm-mission-document.pdf while the study can be downloaded from https://www.researchgate.net/publication/322244341_How_Does_Government_Microfinance_Impact_the_Rural_Poor_Evidence_from_Madhya_Pradesh
8. Visit the website https://www.ifad.org/en/web/ioe/ppe—and select one evaluation study and critically analyse the possibility of strengthening the evaluation design by building on the various livelihood frameworks. Two students should not select the same evaluation study. Indicate the IFAD URL of the study in your assignment.
9. We have studied six rural livelihood approaches so far, namely: (1) DFID sustainable livelihoods approach (2) Ostrom's governing the commons (3) Livelihood trajectories (4) Rights-based approach (5) BRAC graduation approach and the (6) Resilience approach. Discuss how the application of these approaches adds value to the situations that researchers and development professionals face in their day-to-day functioning.
10. Tambe et al. (2017) while evaluating a public microfinance programme in central India assessed the investment decisions that households took in asset creation, basic needs, education/health, and

social needs along the socio-economic gradient. Based on the data provided below assess the livelihood strategy along the gradient of poverty by using the lens of the various rural livelihood approaches.

Investment	Very poor	Poor	Medium	Well off	Total
Asset creation	**37%**	**63%**	**70%**	**63%**	**62%**
Natural assets	62%	65%	69%	87%	69%
Physical assets	38%	33%	31%	13%	31%
Basic needs	**35%**	**19%**	**19%**	**23%**	**21%**
Debt repayment	69%	26%	26%	9%	29%
Food and related	17%	32%	32%	22%	29%
Housing	14%	37%	11%	43%	33%
Others	3%	5%	32%	30%	10%
Education/health	**19%**	**12%**	**7%**	**8%**	**12%**
Education	47%	42%	43%	25%	42%
Health	53%	58%	57%	75%	58%
Social needs	**8%**	**6%**	**4%**	**6%**	**6%**
Marriage	100%	67%	75%	67%	67%
Festival and others	0%	33%	25%	33%	33%
Grand total	**100%**	**100%**	**100%**	**100%**	**100%**

11. Based on a livelihood study, Patnaik et al. (2017) provide a schematic representation of the transformation of the asset pentagon of the households supported under the National Rural Livelihood Mission (NRLM), a public microfinance programme. Analyse how the use of the asset pentagon helped in providing a deeper understanding and visualization of the project impacts.

2011: Baseline scenario

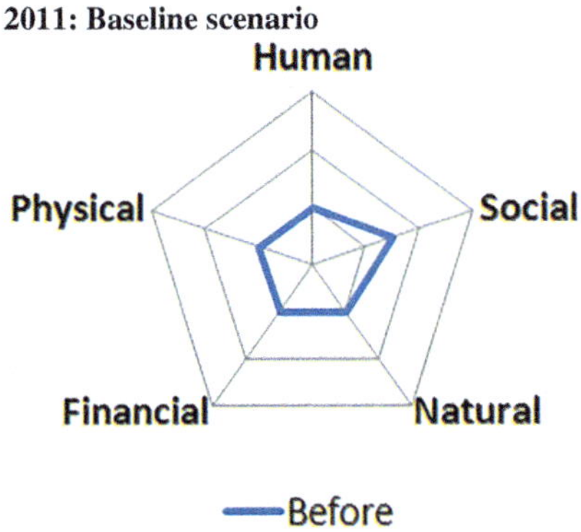

The households targeted under the NRLM programme suffer from multiple vulnerabilities. Being mostly illiterate, landless, poor, and assetless, with unskilled labour work being their main source of income, they are caught in a poverty trap. Consequently, their asset pentagon is depressed. To cope with stress and shocks, they do not have access to social safety nets to prevent them from falling deeper into poverty. Of the three elements of social capital, namely bonding, bridging, and linking, only bonding element was present between the neighbourhood groups

2012: SHG and VO formation

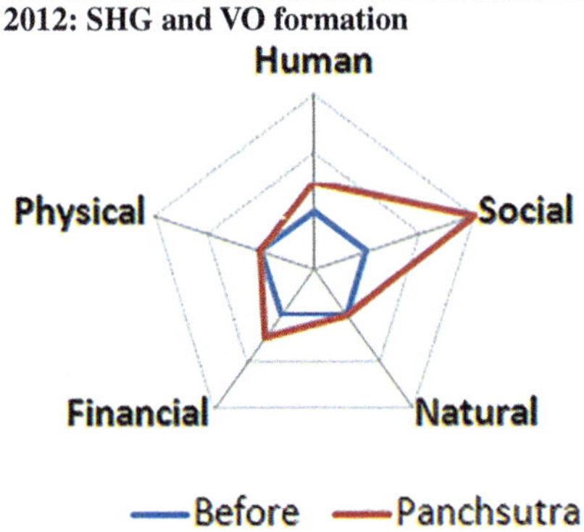

The women were organized into Self Help Groups (SHGs) and a behaviour comprising of five elements (*panchsutra*) regular saving, regular lending, regular meetings, regular repayment, and record keeping was embedded in them. The bridging element of the social capital is the sharing and exchange between the groups which was provided by forming the Village Organization by federating the SHGs at the village level. The linking element of the social capital is the relationship with the higher authorities, such as government departments, which was enabled with the association with NRLM. This led to a rapid expansion of the social capital which provided a platform for accumulating the other assets.

<table>
<tr>
<td>2015: Access to credit (well off)
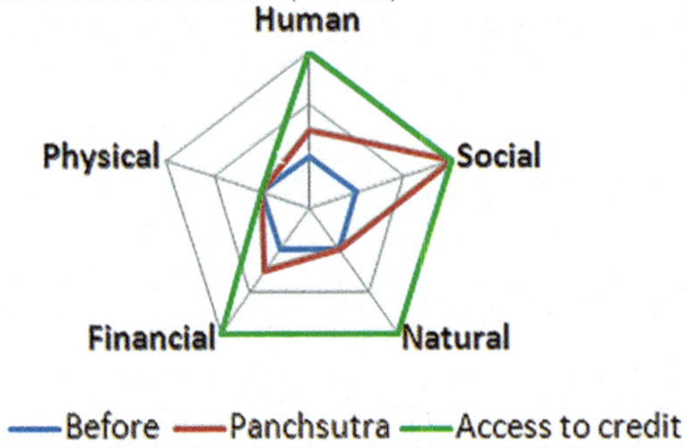
</td>
<td rowspan="2">Access to credit in the form of revolving fund, community investment fund, and bank loans was used by the households to enhance their natural and physical capital. Natural capital investments were by way of taking up on-farm activities such as seeds, fertilizers, vegetable farming, etc., and off-farm activities such as rearing farm animals (cows, goats, pigs, poultry, etc.). Non-farm activities such as kirana shop, tea shop, utensils cart, repair shop, etc. were also set up by the households to enhance the physical capital. The well off strengthened their natural capital, while the others strengthened both their natural and physical capital.
Income from these activities resulted in an accumulation of financial capital, and the resultant investment in health and education further strengthened the human capital.</td>
</tr>
<tr>
<td>2015: Access to credit (very poor, poor, and medium)
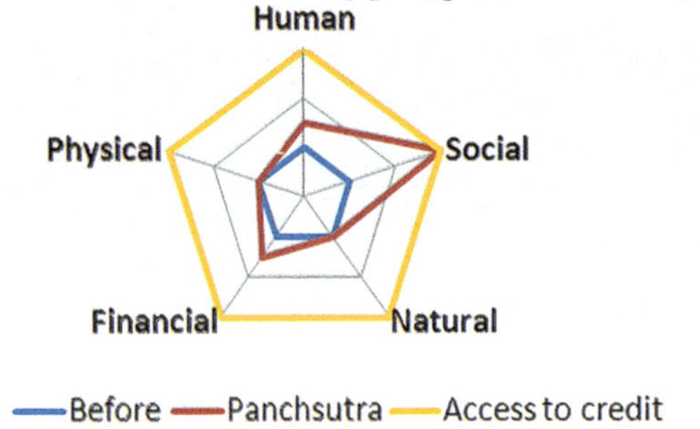
</td>
</tr>
</table>

References

Cash, D. W., Clark, W. C., Alcock, F., Dickson, N. M., Eckley, N., Guston, D. H., Jager, J., & Mitchell, R. B. (2003). Knowledge systems for sustainable development. *Proceedings of the National Academy of Sciences, 100*, 8086–8091. http://crossasia-repository.ub.uni-heidelberg.de/3789/1/India%20Recasting%20Culture.pdf. https://documents1.worldbank.org/curated/en/298391515516507115/pdf/122548-WP-P090764-PUBLIC-India-BRLP-Booklet-p.pdf.

Castillo, G. E. (2008). *The right to a sustainable rural livelihood: Strategies, lessons learned and actions (2008–2016)*. Oxfam novib, The Hague, Netherlands. Retrieved June 24, 2021 at 22:00, from https://oxfamilibrary.openrepository.com/bitstream/handle/10546/316470/bp-right-sustainable-rural-livelihood-120308-en.pdf?sequence=1.

Donohue, C., & Biggs, E. (2015). Monitoring socio-environmental change for sustainable development: Developing a Multidimensional Livelihoods Index (MLI). *Applied Geography, 62*, 391–403.

Ellis, F., & Freeman, H. A. (2004). Rural livelihoods and poverty reduction strategies in four African countries. *Journal of Development Studies, 40*(4), 1–30.

Krishna, A. (2007). For reducing poverty faster: Target reasons before people. *World Development, 35*(11), 1947–1960.

Masud, M. M., Kari, F., Yahaya, S. R. B., & Al-Amin, A. Q. (2016). Livelihood assets and vulnerability context of marine park community development in Malaysia. *Social Indicators Research, 125*(3), 771–792.

Matin, I., & Hulme, D. (2003). Programs for the poorest: Learning from the IGVGD program in Bangladesh. *World Development, 31*(3), 647–665.

OECD. (2007). *Promoting pro-poor growth: Policy guidance for donors*. DAC Guidelines and Reference Series. OECD, MA, USA. Retrieved June 12, 2021 at 15:00, from https://www.oecd-ilibrary.org/docserver/9789264024786-en.pdf?expires=1623490443&id=id&accname=guest&checksum=26CD48280163A39367AF1C9B77771BDD.

Patnaik, S. Tambe, S., Bisaria, J., Pandey, A., Sarma, U. K., & Ananda Vadivelu G. (2017). *Evaluation of Madhya Pradesh Deendayal Antodaya Yojana—State Rural Livelihoods Mission (MPDAY-SRLM)*. Study Report, Indian Institute of Forest Management (IIFM), Bhopal, p. 205. ISBN No. 978-81-943801-1-5.

Qian, C., Sasaki, N., Jourdain, D., Kim, S. M., & Shivakoti, P. G. (2017). Local livelihood under different governances of tourism development in China—A case study of Huangshan mountain area. *Tourism Management, 61*, 221–233.

Scoones, I. (2015). *Sustainable livelihoods and rural development*. Fernwood Publishing.

Tambe, S., Arrawatia, M. L., Bhutia, N. T., & Swaroop, B. (2011). Rapid, cost-effective and high resolution assessment of climate-related vulnerability of rural communities of Sikkim Himalaya, India. *Current Science*, 165–173.

Tambe, S., Patnaik, S., Bisaria, J., Pandey, A., Sarma, U. K., & Ananda Vadivelu, G. (2017). How does Government Microfinance Impact the Rural Poor? Evidence from Madhya Pradesh. *Economic and Political Weekly, LII*(52), 81–87.

Tambe, S., Rawat, G. S., Bhutia, N. T., Sherpa, P. N., Dhakal, S., Pradhan, S., et al. (2019). Building sustainability in the Eastern Himalaya: Linking evidence to action. *Environment, Development and Sustainability, 22*(6), 5887–5903.

Thanh, H. T., Tschakert, P., & Hipsey, M. R. (2020). Moving up or going under? Differential livelihood trajectories in coastal communities in Vietnam. *World Development, 138*, 105219.

Chapter 12
Mainstreaming Livelihood Approaches in Existing Practice

Abstract In this chapter, we explore the value that livelihood approaches can add to livelihood development. The existing practice of development comprising of project planning, monitoring, and evaluation relies on standard project management frameworks. In this chapter, we provide guidance on how rural livelihood approaches can add value to the logical framework approach, theory of change, and monitoring and evaluation frameworks. These generalist project management frameworks that are sector neutral are mandated by funding agencies and hence their use is widespread. We opine that a deeper understanding of livelihoods is ingrained in the rural livelihood approaches that can be harnessed by embedding them while designing, executing, and evaluating livelihood interventions. In this chapter, we explain the 'how' part of integrating rural livelihood approaches using real-life examples. Supplementing the standard management frameworks with these livelihood approaches can result in mainstreaming their use in development. We opine that integrating livelihood approaches in practice can help in improving the efficacy and efficiency of development programmes.

Keywords Logical framework approach · Situation analysis · Problem analysis · Objective analysis · Sequencing · Theory of change · Monitoring · Evaluation · Integrating · Key performance indicators · Baseline · Endline

12.1 Introduction

In this chapter, we explore how rural livelihood approaches can sharpen livelihood practice in the sectoral area of poverty, livelihoods, and rural development. In development, various donors, bilateral, and multilateral agencies prescribe the use of generalistic project management frameworks as these are sector neutral and can be applied across diverse sectors such as health, education, environment, sanitation, water, agriculture, forests, livelihoods, etc. This has led to the proliferation of generalist frameworks such as logical framework approach, theory of change, and others which are extensively used in development. The RLAs can add immense value to these project management frameworks in planning, execution, monitoring, and evaluation as we detail below:

12.2 Integrating RLAs in Project Planning

In project planning, the two frameworks that are widely used are the logical framework approach (LFA) and the theory of change (TOC). We discuss the integration of the rural livelihood approaches in these project management frameworks and what value they can add.

S. Tambe, *Teaching and Learning Rural Livelihoods*, Sustainable Development Goals Series,
https://doi.org/10.1007/978-3-030-90491-3_12

12.2.1 Adding Value to the Logical Framework Approach (LFA)

The Logical Framework Approach (LFA) is a generalist framework that provides a structured and logical sequence of steps for designing projects and their monitoring and evaluation (Box 12.1). It comprises two phases—the analysis phase and the planning phase. The analysis phase results in the formulation of the project strategy and comprises four tools, namely stakeholder analysis, problem analysis, objective analysis, and strategy analysis. The planning phase develops the logframe matrix, activity scheduling, and resource scheduling. The RLAs can be embedded in the various stages of the LFA to add value as shown in Table 12.1. A practical example of how the RLAs can add value to an LFA of a rural livelihoods project is shown in Box 12.2. In this example, we show how the livelihood trajectory approach can help discern the best practices and home-grown success stories for the project intervention to build on and strengthen its relevance. Similarly, the eight design principles of Ostrom's common pool resources (CPRs) can help in establishing a democratic forest governance mechanism. The step-by-step graduation approach can help in sequencing the inputs to the ultra-poor households so that they escape poverty in a time-bound manner.

Box 12.1 Logical Framework Approach

The US Agency for International Development (USAID) had conceived and designed the Logical Framework Approach (LFA) in 1969 to address the three concerns about project management. These concerns included vague planning without clearly defined objectives, undefined management responsibilities, and the adversarial nature of the evaluation as there would often be disagreement on the intended project outcomes.

LFA is an analytical process and a set of objective tools that support and inform project management. It consists of a set of interlocking concepts that aid in the systematic analysis of a problem or an undesired situation to evolve a better understanding of the project rationale, its intended objectives, the means to achieve it, and the measurement. The approach helps in improving the planning, monitoring, and evaluation of a development intervention. It has been extensively embedded in the regular functioning of most donor agencies, multilaterals, and bilateral development agencies. The framework is divided into two phases: the analysis phase and the planning phase. Each of these phases is further comprised of a sequential set of objective tools, namely: stakeholder analysis, situation analysis, problem analysis, objective analysis, strategy analysis, developing logframe, activity scheduling, and resource scheduling.

Source: Örtengrena (2004), UNDP (2009), European Integration Office (2011)

Box 12.2 Mainstreaming RLA in LFA of a Rural Livelihoods Project

The Logical Framework Approach (LFA) of this IFAD funded tribal development project in central India is provided on Page 31–34 of the project impact evaluation document available at: https://www.ifad.org/documents/38714182/39737274/Impact+Evaluation+of+the+Jharkhand+%E2%80%93+Chhattisgarh+Tribal+Development+Programme+%28JCTDP%29.pdf/758ced1c-5497-41c7-bea5-c3a4906b0e6d

How do you think embedding RLA can add value to this LFA?

We demonstrate this for Outputs 4, 5, and 6 as detailed below:

Output 4: Appropriate farming system technologies identified, developed and adapted, tested with poor farmers, and made widely available.

- The livelihood trajectory approach can be used to identify households that have escaped poverty, along with their livelihood pathways to ascertain the best practices and home-grown success stories. Similarly, households that have fallen back into poverty can provide insights into the gaps in the safety net programs that need to be bridged. Thus, the pathways into and out of poverty can inform the appropriate protection, prevention, and promotion schemes to be adopted.

Output 5: Tribal rights on natural resources such as land, forest, water, minor minerals, recognized and promoted.

- Ostrom's eight design principles of common pool resources (CPRs) can be built upon to nurture a common property regime. These design principles can provide the building blocks such as the need for clear resource boundaries, laying down the access and benefit sharing arrangements, local enforcement mechanism, resource monitoring, and recognition by external government authorities.
- Rights-based approaches can aid in a power analysis to identify the stakeholders who wield power and control these resources and the ones who are discriminated against.

Output 6: Complementary income-generating, expenditure saving, and viable microenterprise, benefiting especially the 'losers', PTGs, and women, in forest, farm, and off-farm sectors, promoted and implemented in programme villages.

- Instead of the 'loser household', the 'ultra-poor household' terminology can be used as it is a better defined and standard category. The ultra-poor households can be objectively identified using the targeting methodology of the graduation approach.
- Before jumping directly to incomes and microenterprise, 'consumption support' and 'savings' need to be provisioned for the ultra-poor/loser households. This will ensure that their basic needs are first met, liberating them from the stress of day-to-day survival, thereby enabling them to focus on livelihood opportunities and planning for the future. Savings habit will develop financial discipline and ensure that they do not have to sell off their assets when faced with ill-health or a break in work opportunities. A step by step, sequential set of inputs is needed to lift the ultra-poor out of poverty and to help them graduate to microenterprise and sustainable livelihoods.
- The livelihood strategy needs to build on the asset base of the ultra-poor households, their vulnerability context, and the mediating policy, institutions, and processes.

Table 12.1 Adding value to the Logical Framework Approach (Adapted from European Integration Office, 2011)

Phase	LFA stage	Adding value by embedding RLAs
Analysis	▼**Stakeholder analysis:** Identifies the stakeholders, their characteristics, their problems, interests, potentials, deficiencies, their relations, and implications for the project	• SLA can assist in identifying the institutions and processes and in assessing the role they play in influencing livelihood outcomes • Graduation approach emphasizes the heterogeneity amongst the poor and the need to treat the ultra-poor as a separate stakeholder category • Assessing the CPR regimes using the design principles of Ostrom (Box 5.2) • A rights-based approach can help ascertain the distribution of power, by identifying groups lacking effective rights and those that may be denying rights to others
	▼**Situation and problem analysis:** Identifies undesired situation, constraints, and opportunities, determining cause and effect relationships	• SLA can assist in highlighting the vulnerability context, deficiencies in the asset pentagon, and the role of ongoing programmes, policies, and institutions • Resilience framework can help in assessing the risks and shocks to livelihoods • Livelihood trajectory can provide insights on how the households have progressed over the last few decades, the drivers behind their escape or descent into poverty, and the performance of protection, prevention, and promotion schemes • The status of the CPRs and their contribution to provisioning livelihoods for the poor can be studied • Rights-based approach can ascertain if the root cause of the problem stems from structural, systemic rights barriers to people's empowerment, capacity-building, and government accountability
	▼**Objective analysis:** Developing solutions for the identified problems; identifying means to end relationships, imagining an improved situation in the future	• Livelihood strategy can build on the three elements of SLA, namely the vulnerability context, asset mix, and the prevailing policies, institutions, and processes • Livelihood trajectory approach can provide insights into which livelihood strategies are working in the local context • Rights-based approach ensures focus on marginalized and vulnerable groups with the outcome of changing behaviour of right holders and duty bearers • Graduation approach provides guidance on a step-by-step, sequential approach of targeting the ultra-poor, consumption support, savings, market analysis of feasible livelihood options, skill transfer, asset transfer, access to credit, and continuous life coaching to lift the ultra-poor out of poverty • Resilience framework can help in designing interventions that help in protecting the asset base in times of shock
	▼**Strategy analysis:** involves the identification of possible alternatives	Not applicable, as this step is specific to the scope of the project. Based on the objective and mandate of the project, a subset of the project objectives can be selected from amongst the proposed interventions

(continued)

Table 12.1 (continued)

Phase	LFA stage	Adding value by embedding RLAs
Planning	▼**Developing logframe:** Defining project structure, testing its internal logic and risks, and formulating measurable indicators of success. It often requires further analysis and refinement of ideas	• SLA can assist in formulating the outcomes of livelihood projects and their objectively verifiable indicators. The asset pentagon can also be monitored to assess the long-term impacts of the project • The SLA and resilience framework can help in framing the risks and assumptions to the project outputs and outcomes • Livelihood trajectory approach can assist in assessing the livelihood outcomes and their attribution to the project interventions • The conservation status of the CPR and its contribution to the livelihoods of the poor can be constructed as one of the outcomes • The rights-based approach will require activities like research, advocacy, evaluation, public education, and organizational development • The resilience framework can help in assessing the sustainability of the intervention in the face of shocks and stresses
	▼**Activity scheduling:** Determining activities, their sequence, schedule, and dependency, estimating their duration, and assigning responsibilities	• Ensure transparency, consultation, access to information, participation, and grievance redressal to strengthen accountability • Graduation approach prescribes a sequenced set of activities for lifting the ultra-poor households out of poverty, which the activity scheduling can take guidance from
	▼**Resource scheduling:** From the activity schedule, developing resource requirements, input schedules, and budget	• Not applicable, as is project specific involving budget and staff allocation to specific activities

12.2.2 *Enriching the Theory of Change (TOC)*

The Theory of Change is designed to enhance the capacity of development organizations to achieve their goals and social impact (Box 12.3). It is both a tool to map out the logical sequence of an initiative from inputs to outputs to outcomes, as well as a deeper reflective process on how and why change might happen as an outcome of the initiative (Vogel, 2012). At the heart of a theory of change is a power analysis and a change hypothesis. In power analysis, we assess what is the key change we desire, what are the forces that are driving or blocking such a change, and which stakeholders are likely to play a key role as allies or opponents (Green, 2013). Is this power formal or informal, visible or invisible or hidden (Green, 2013)? Power analysis helps in the formulation of a change hypothesis as to how the change will come about. The rights-based-approach can guide this analysis of power relations and help in unearthing structural inequalities. Much of the value that RLAs can add to LFA, holds for TOC as well. The SLA can guide in preparing the TOC based on the three elements of vulnerability context, asset pentagon, and the macro context. It can help assess the mediation role of policies, institutions, and processes (PIPs) in converting the assets into outputs and outcomes. The graduation approach provides a sequenced set of interventions for driving socio-economic change at the household level. The livelihood trajectory approach can assist in clarifying power structure, impact pathways, and discern causality. An applied example of how the RLAs can add value to a TOC is described in Box 12.4.

Box 12.3 Theory of Change

Theory of Change (ToC) is a methodology that is widely used in planning, participation, and evaluation at companies, philanthropy, not-for-profit, and government sectors to promote social change. It defines long-term goals and helps in mapping them backward to identify necessary preconditions. Once a long-term goal is identified, the project proponents then determine the conditions that must be in place such that the goal is reached. Any such necessary conditions should be projected as outcomes on the Theory of Change pathway, underneath the long-term outcome. These outcomes then act as preconditions to the long-term outcome. The process of determining preconditions continues, drilling down the pathway by exploring fundamental questions such as: 'What has to be in place for an outcome to be achieved?' and 'Are the preconditions adequate for the outcome to be achieved?' In these sessions, participants may use markers, sticky notes, and chart paper to identify and organize outcomes, surface assumptions, develop indicators, and so on. The innovation of the Theory of Change is in making the distinction between desired and actual outcomes and in needing the stakeholders to model their desired outcomes before they choose forms of intervention to achieve those outcomes.

Source: Wikipedia, retrieved on 31st May 2021

Box 12.4 Mainstreaming RLA in TOC of a Livelihood Project

The Theory of Change (TOC) of a rural livelihood project in Bihar, in the eastern part of India, is provided on Page 13 of the project design document (Hoffman et al., 2018) available at: http://www.3ieimpact.org/sites/default/files/2019-01/IE71-Bihar-livelihoods.pdf

How do you think embedding RLA can add value to this TOC?

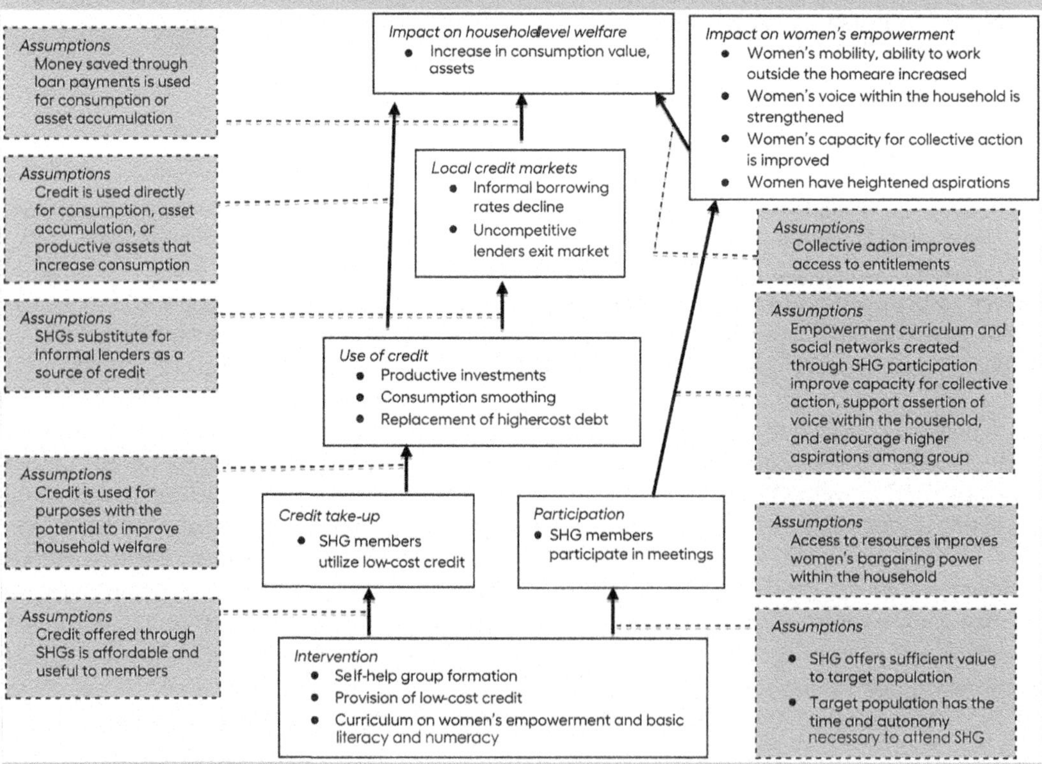

Box 12.4 (continued)

We explore the value that RLAs can add to this TOC by applying them in this context:

Graduation approach: The TOC assumes the SHG members to be homogenous and that they will respond uniformly. In reality, the SHG members belong to different socio-economic categories and will consequently follow diverse pathways. In the absence of consumption support prescribed by the graduation approach, the ultra-poor will mostly utilize the credit for consumption smoothening and other non-productive investments such as health, food, clothing, house repair, retiring debts, etc. and hence will face difficulty in repaying the SHG credit and may turn into defaulters. The poor and middle-class households from amongst the SHG members will crowd out the ultra-poor from accessing loans when credit supply is limited.

Sustainable livelihoods approach: The vulnerability context and asset pentagon of the SHG households will determine how they respond or benefit from the SHG intervention. The change in the asset pentagon can be used as a key performance indicator.

Rights-based approach: Can guide how the SHG intervention can bring about a structural change in the external environment related to exclusion, discrimination, and lack of accountability and thereby contribute to women's empowerment.

Resilience framework: Will assess the capacity of the SHG members to maintain their livelihood status in the face of shocks and stresses.

Livelihood trajectories: The livelihood trajectory of the SHG members can be classified through a livelihood strategy typology as 'dropping out', 'hanging in', 'stepping up', and 'stepping out' to gain a deeper understanding of the programme impacts.

Hence, we note that the livelihood approaches can add value in terms of providing guidance on the heterogeneity of the target group, their varied response patterns to the developmental intervention, the structural changes that will be needed to empower them, and how their diverse livelihood trajectories can be classified.

12.3 Integrate RLAs in Project Monitoring and Evaluation

In project monitoring and evaluation, the IFAD five-pronged evaluation approach (IFAD, 2009) and the Double Difference Approach (Gertler et al., 2016) are widely used, and both can benefit from the integration of the rural livelihood approaches by developing the baseline and endline, constructing the key performance indicators, and assessing sustainability of the livelihood outcomes (Table 12.2). The IFAD evaluation framework is based on the five components of relevance, effectiveness, impact, efficiency, and sustainability (Box 12.5). Various livelihood approaches can inform the IFAD evaluation framework and sharpen it. The livelihood trajectory approach can inform the relevance of developmental interventions by discerning existing livelihood pathways in practice. The graduation approach offers a whole portfolio of interventions that can be mapped to the socio-economic category of households thereby aiding in investment efficiency. An example of how this approach can be embedded in an evaluation study of a rural livelihoods programme is provided in Box 12.6. The change in the asset pentagon and vulnerability context can help in measuring programme effectiveness as well as impact. The livelihood resilience framework can assess the sustainability of livelihood development. The double difference approach or the difference-in-differences approach is a tool for impact evaluation that compares the changes in the outcome of interest over time between the treatment group and the comparison group as described in Box 12.7. The focus is on ascertaining the impact that is the changes

Table 12.2 Adding value to monitoring and evaluation

Stage	RLA value addition
Assessing the baseline and endline	• Livelihood trajectory approach can help in assessing the percentage of households in the various wellbeing categories before and after the project to construct the transition matrix which can be a very useful village-level performance indicator. This can be followed by qualitatively assessing the attribution of the project to the livelihood outcomes by documenting the household life stories
Constructing the Key Performance Indicators (KPI)	• Programme outcomes and their objectively verifiable indicators can be constructed with guidance from the two elements of SLA, namely asset pentagon and livelihood outcomes. • Graduation criteria along with the literature on post-graduation livelihood trajectories can also assist in constructing the key performance indicators (KPIs)
Assessing sustainability of the outcomes	• Sustainability of livelihoods in the face of shocks and stresses can be assessed using the resilience framework and the SLA with its vulnerability context and asset pentagon

Box 12.5 IFAD Five-Pronged Evaluation Framework

The five-pronged evaluation criteria of IFAD comprised of:

1. **Relevance** concerns the extent to which a development initiative and its intended outputs or outcomes are consistent with the needs of the environment and the intended beneficiaries. In applying the criterion of relevance, evaluations should explore the extent to which the planning, design, and implementation of initiatives take into account the local context.
2. **Effectiveness** is a measure of the extent to which the initiative's intended results (outputs or outcomes) have been achieved or the extent to which progress toward outputs or outcomes has been achieved. Evaluating effectiveness in project evaluations involves an assessment of cause and effect, that is, attributing observed changes to project activities and outputs. Assessing effectiveness involves two basic steps. Measuring the change in the observed output or outcome and attributing the observed change to the initiative.
3. **Impact** measures change in human development and people's wellbeing that are brought about by development initiatives, directly or indirectly, intended, or unintended. Confirming whether benefits to beneficiaries can be directly attributed to the intervention can be difficult, especially since there are several ongoing interventions often with overlapping objectives.
4. **Efficiency** includes a measure of how economically inputs (funds, expertise, time, etc.) are converted into results. An analysis of budget use and compliance with timetables is also important to assess the efficiency dimension.
5. **Sustainability** concerns the likely continuation of net benefits from an intervention beyond the phase of funding support. It also includes an assessment of the likelihood that actual and anticipated results will be resilient to risks beyond the project's life.

Source: IFAD (2009)

that can be directly attributed to the programme (Gertler et al., 2016). The RLA can assist in the assessment of baseline, endline, livelihood trajectory, and asset pentagon of both the treatment group and the counterfactual.

Box 12.6 Integrating BRAC Graduation Approach in Evaluation

In 2011, the Government of India launched the National Rural Livelihood Mission (NRLM) to reduce rural poverty by providing access to gainful self-employment and skilled wage employment. This programme aimed to restructure the earlier avatar that suffered from a lack of mobilization of the beneficiaries, patchy quality of SHGs, and weak bank linkage. In 2016, after 5 years a concurrent evaluation was carried out in the central Indian state of Madhya Pradesh—a front runner in the implementation of this programme.

In terms of targeting performance, the evaluation found that programme beneficiaries were landless or marginal farmers, functionally illiterate, belonged to socially marginalized groups, were assetless, and were compelled to migrate in the agriculture lean season to work as casual labour. During times of shock or distress, they had no option but to borrow money from the local money lenders at high interest rates. They lived in fragile huts, were unaware of their entitlements and government schemes hardly reached them.

In terms of social impact, the evaluation found that the role of the woman which was earlier limited to the household expanded rapidly. Income-generating activities started revolving around her, as she became the source of credit. The SHG gave her a new, common identity '*samuha ki didi*' (SHG sister) which helped diminish her other identities related to caste, class, and religion. The women started participating in household decision making, in development programmes, and in rooting out social evils. They not only joined the mainstream but also became agents of change, prosperity, and social transformation.

The evaluation found that the economic impact of the credit uptake differed for the ultra-poor and poor. The ultra-poor who constituted 7% of the project households prioritized basic survival and social needs such as food security, debt repayment, health expenditure, and marriage expenses. Hence, they could invest only 37% of the micro-loans on income-generating assets and consequently, a sizable percent of the loans taken by them were outstanding. On the contrary, most of the poor households who constituted the bulk (64%) of the project households had diversified their asset portfolio by investing 63% of their loans in asset building. Consequently, in 5 years, while the poor had graduated to the vulnerable non-poor category, the ultra-poor were still stagnating in poverty and finding it difficult to repay the loans.

The graduation approach puts the spotlight on the ultra-poor. By treating them as a separate category, it seeks to better understand their differential response to developmental interventions compared to the poor. Without this level of granularity, the ultra-poor who are the poorest would have been left behind.

Source: Tambe et al. (2017), Patnaik et al. (2017)

Box 12.7 Double Difference Approach

The double difference approach, also called the difference-in-difference is a quantitative research approach in social sciences that seeks to mimic an experimental research design using observational data by studying the differential effect of a treatment on a 'treatment group' and compare it with a 'control group' in a natural experiment. It seeks to calculate the resultant effect of a treatment (an explanatory variable or an independent variable) on an outcome (a response variable or dependent variable) by comparing the average change over time in the outcome variable for the treatment group and the average change over time for the control group.

Source: Wikipedia, retrieved on 31st May 2021

12.4 The Way Forward

While the use of generalist project management frameworks has proliferated, they provide an opportunity to embed the rural livelihood approaches, thereby mainstreaming them in development practice. In the sectoral area of rural livelihoods, poverty alleviation, and rural development, the integration of the rural livelihood approaches can help sharpen the generalist frameworks used in project planning, monitoring, and evaluation. These generalist frameworks are mandated by the funding and development agencies and provide an opportunity for the development professionals to embed the rural livelihood approaches in them. This integration will generate a nuanced understanding of rural livelihoods compared to the stand-alone use of these generalist frameworks. We opine that RLAs bring with them a deeper understanding of rural livelihoods, poverty alleviation, and rural development, which can immensely benefit development practice by embedding these approaches in projects. The marriage of these two disciplines, namely rural livelihood approaches and project management can result in a win-win situation with better designed projects delivering greater livelihood impact.

Additional Resources

1. **LFA resource material**
   - European Integration Office (2011). Guide to the Logical Framework Approach http://www.evropa.gov.rs/Evropa/ShowDocument.aspx?Type=Home&Id=525
   - Örtengrena (2004). A summary of the theory behind the LFA method https://cdia.asia/wp-content/uploads/2014/09/SIDA_LFA.pdf
   - Team Technologies (2005). The logframe handbook: A logical framework approach to project cycle management http://www.evropa.gov.rs/Evropa/ShowDocument.aspx?Type=Home&Id=525
2. **TOC resource material**
   - Compilation of useful resource material https://www.researchtoaction.org/2011/05/theory-of-change-useful-resources/
   - Using TOC in evaluation https://www.betterevaluation.org/en/managers_guide/step_2/describe_theory_of_change
   - Examples of TOC https://assets.publishing.service.gov.uk/media/57a08a66ed915d622c000703/Appendix_3_ToC_Examples.pdf
3. **IFAD evaluation manual** http://www.ifad.org/evaluation/process_methodology/doc/manual.pdf
4. **World Bank double difference approach** https://www.worldbank.org/en/programs/sief-trust-fund/publication/impact-evaluation-in-practice

Exercises

1. The evaluation of a 5-year sustainable rural livelihoods project reveals the following. Indicate on which of the five components (relevance, effectiveness, impact, efficiency, and sustainability) of the IFAD Evaluation Framework did the project fail? Also, indicate the integration of which RLAs could have possibly prevented these lapses?
   (a) The greenhouses constructed for the households for promoting vegetable farming were not used as this livelihood was alien to this herding village traditionally rearing yaks for a living
   (b) The ultra-poor were excluded while identifying the beneficiary households
   (c) The water tanks constructed for drinking purpose developed a leak in the first year itself
   (d) The tourists stopped visiting the ecotourism site from the sixth year as an alternative route got developed
   (e) The farmers could not get a fair price for the pigs they had reared

(f) The tissue culture Eucalyptus clones procured from a private nursery performed well but the same clones were available at a much cheaper price in the government nursery
(g) The cows distributed were sold off by the beneficiary households as they had no experience in dairy farming
(h) The survival percentage of the horticulture plants dropped off after the fifth year
(i) The construction of the minor irrigation channel got delayed by 6 months
(j) With the money earned from goat rearing, the households indulged in gambling and drinking

2. In addition to reporting on 'what' the performance was, evaluation needs to wherever possible strive for a deeper understanding of 'why' the performance was as it was. Discuss how the various RLAs can be embedded in the evaluation to strengthen the attribution of the impacts to the project interventions.
3. Study the evaluation of the project titled 'Livelihoods Improvement Project in the Himalayas' funded by IFAD. Analyse how this performance assessment could have been made more robust by embedding rural livelihood approaches in the design. https://www.ifad.org/documents/38714182/39733776/liph_ppa.pdf/52c9604c-eba1-4cef-a04b-885a5e35de7a
4. Study the Theory of Change (TOC) of the 'Rural Livelihood Improvement Programme in Cambodia' funded by IFAD provided for in Annexure-I of this document. Discuss how the rural livelihood approaches can be integrated with this TOC to make it more robust. https://www.ifad.org/documents/38714182/41815811/Approach+Paper+-+Kingdom+of+Cambodia+-+Rural+Livelihoods+Improvement+Project+-+Project+Performance+Evaluation.pdf/0e845aff-6137-3846-d832-11a69eebc96e
5. Study the Theory of Change (TOC) of the 'Tejaswini Rural Women's Empowerment Programme in India' funded by IFAD provided for in Annexure-VII of this evaluation report. Discuss how the rural livelihood approaches can be integrated with the TOC to make it more robust. https://www.ifad.org/documents/38714182/39733776/India+PPE_FINAL.pdf/7c4a0934-5ae1-6284-ee2b-5a1981dc9e59
6. Study the reconstructed Theory of Change (TOC) of the 'Participatory Natural Resource Management Programme in Palestine' funded by IFAD provided for in Annexure-VI of this evaluation report. Discuss how the TOC can be improved by embedding the learnings from rural livelihood approaches. https://www.ifad.org/documents/38714182/40811091/Palestine+PPE_for+web.pdf/57a472a3-05a3-4356-a79b-074aa55f1b24
7. Study the Theory of Change (TOC) of the project titled 'Root and Tuber Improvement and Marketing Programme in Ghana' funded by IFAD provided for in Annexure-VI of this evaluation report. Discuss how the TOC can be improved by embedding the learnings from rural livelihood approaches. https://www.ifad.org/documents/38714182/40811021/ghana_1312_2018.pdf/6e2a36e1-c84b-452d-9747-d84ac8d6c27a
8. Study the proposal and Theory of Change (TOC) of the project titled 'Combating land degradation through integrated and sustainable range and livestock management to promote resilient livelihoods in Northern Punjab' funded by GEF. Discuss how the TOC can be strengthened by integrating the rural livelihood approaches in the project document. https://publicpartnershipdata.azureedge.net/gef/GEFProjectVersions/3b729c71-b9f8-ea11-a815-000d3a337c9e_PIF.pdf
9. Study the participatory wealth ranking (PWR) provided on page 24 of the IFAD 'how to do notes' series titled 'poverty targeting, gender equality and empowerment during project design' and discuss how it builds on the RLA, and what further aspects can be included. https://www.ifad.org/documents/38714170/41240300/How+to+do+note+Poverty+targenting%2C+gender+equality+and+empowerment+during+project+design.pdf/0171dde5-e157-4a6a-8e00-a2cafaa0e314
10. Study the project design and impact of the World Bank funded Bihar Rural Livelihoods Project (Jeevika) of India. Discuss how the integration of the rural livelihood approaches could have

benefited the project. https://projects.worldbank.org/en/projects-operations/project-detail/P090764. https://documents1.worldbank.org/curated/en/298391515516507115/pdf/122548-WP-P090764-PUBLIC-India-BRLP-Booklet-p.pdf

11. Study the two guidance documents prepared by the Swiss Development Corporation (SDC) on how they apply the rights-based approach to the results based management (RBM) and project cycle management (PCM). Discuss the value that the rights-based approach can add to these two project management frameworks in the context of development projects. .https://www.shareweb.ch/site/Conflict-and-Human-Rights/Documents/Policy%20Brief%20-%20CAPEX%20HRBA%20and%20Results%20Measurement.pdf. https://www.shareweb.ch/site/Conflict-and-Human-Rights/Documents/HRBA%20to%20PCM%20%2806.11.2020%29.pdf
12. Institutional analysis for rural development focusses on how institutions influence poverty and the livelihoods of poor people. IFAD has developed guidance notes for 'Institutional Analysis in Rural Development Programmes' that provide a framework and analytical tools for designing programmes and projects based on the core principles of good governance, focussing on pro-poor governance, and systemic sustainability. Discuss how these guidelines integrate the rural livelihood concepts of asset pentagon, vulnerability context, power and processes, institutional context, and livelihood outcomes. https://www.ifad.org/documents/38714170/39144386/Guidance+Notes+for+institutional+analysis+in+rural+development+programmes+-+an+overview.pdf/20cd8002-6c4c-4a83-9e14-034beb66c3ef

References

European Integration Office. (2011). *Guide to the logical framework approach, Republic of Serbia*. European Integration Office, Belgrade, Serbia. Retrieved May 31, 2021 at 19:40. from http://www.evropa.gov.rs/Evropa/ShowDocument.aspx?Type=Home&Id=525.

Gertler, P. J., Martinez, S., Premand, P., Rawlings, L. B., & Vermeersch, C. M. (2016). *Impact evaluation in practice* (2nd ed.). World Bank. Retrieved July 3, 2021 at 16:00, from https://www.worldbank.org/en/programs/sief-trust-fund/publication/impact-evaluation-in-practice.

Green, D. (2013). *What is a theory of change and how do we use it?* Oxfam (blog). Retrieved June 2, 2021 at 13:10, from http://www.oxfamblogs.org/fp2p/?p=15532.

Hoffman, V. Rao, V., Datta, U., Sanyal, P., Surendra, V., & Maumdar, S. (2018). *Poverty and empowerment impacts of the Bihar Rural Livelihoods Project in India*. 3ie, Impact Evaluation Report 71. New Delhi: International Initiative for Impact Evaluation (3ie). Retrieved May 26, 2021 at 14:00, from http://www.3ieimpact.org/sites/default/files/2019-01/IE71-Bihar-livelihoods.pdf.

IFAD. (2009). *Evaluation manual: Methodology and processes*. International Fund for Agriculture Development (IFAD). Retrieved May 25, 2021 at 14:00. from http://www.ifad.org/evaluation/process_methodology/doc/manual.pdf.

Örtengrena, K. (2004). *A summary of the theory behind the LFA method*. SIDA, Stockholm. Retrieved May 31, 2021 at 14:30, from https://cdia.asia/wp-content/uploads/2014/09/SIDA_LFA.pdf.

Patnaik, S. Tambe, S., Bisaria, J., Pandey, A., Sarma, U. K., & Ananda Vadivelu G. (2017). *Evaluation of Madhya Pradesh Deendayal Antodaya Yojana—State Rural Livelihoods Mission (MPDAY-SRLM)*. Study Report, Indian Institute of Forest Management (IIFM), Bhopal, p. 205. ISBN No.978-81-943801-1-5.

Tambe, S., Patnaik, S., Bisaria, J., Pandey, A., Sarma, U. K., & Ananda Vadivelu, G. (2017). How does Government Microfinance Impact the Rural Poor? Evidence from Madhya Pradesh (December 30, 2017). *Economic and Political Weekly, LII*(52), 81–87.

Team Technologies. (2005). *The logframe handbook: A logical framework approach to project cycle management (English)*. World Bank Group. Retrieved May 31, 2021 at 19:30, from http://documents.worldbank.org/curated/en/783001468134383368/The-logframe-handbook-a-logical-framework-approach-to-project-cycle-management.

UNDP. (2009). *Handbook on planning, monitoring and evaluating for development results*. United Nations Development Programme. Retrieved May 31, 2021 at 14:00, from http://web.undp.org/evaluation/handbook/documents/english/pme-handbook.pdf.

Vogel, I. (2012). *Review of the use of 'Theory of Change' in international development*. DFID.

Chapter 13
The Way Forward

Abstract This chapter summarizes the evolution of rural development thinking, the conceptual advancements during the MDG period, sustenance of rural areas, emerging issues during the SDG period, and the way forward. Till the 1980s the rural development thinking took on a simplistic focus of a welfare state with the state playing the role of a patron. The thinking evolved with the advent of new ideas that pursued a new-liberal, market-led approach. But both the government and market-led approaches could not address the specific issues of persistent poverty and rising disparity. This led to the growth of normative concerns and rights-based approach that focussed on removing structural inequalities. The MDG period saw impressive achievements in halving poverty, but it also threw up new challenges with middle income countries housing most of the poor, with the poorest residing in low-income countries. The late developing countries in South Asia and SSA grappled with provisioning decent jobs and small holder farming persisted. How to keep rural livelihoods thinking relevant in the future? We opine that a complementary fusion of the six livelihood approaches discussed in this book can help enhance their contribution in achieving the SDG goals of ending poverty and hunger.

Keywords Development thinking · Welfare state · Market-led approach · Participatory methods · Rights-based approach · Social capital · Decent jobs · Livelihoods agenda · Power relations · Planetary processes · Anthropocene · Resilience

13.1 Evolution of Livelihoods and Development Thinking

Over the past five decades, development policy and practice have been shaped by successive waves of ideological thinking that have defined and redefined our understanding of development thinking. The development thinking has evolved from a state-led perspective with the state playing a dominant role, to an actor-based, market-led perspective that looked at the poor as proactive agents of transformation, to the rights-based approach with normative concerns of capability, equity, sustainability, and empowerment, to sustainability that looks at integrating environmental sustainability with development (de Haan & Zoomers, 2005; Bohle, 2009). In the 1950s in the post-colonial era, the understanding was that the state as a patron needs to take the onus of driving rural development (Opschoor et al., 2005). In this welfare state model, rural development thinking was based on the three paradigms of production, employment, and income poverty (Bohle, 2009). These paradigms focussed on resources and manifested themselves in the form of agriculture-led green revolution to enhance production, integrated rural development to provide jobs, and income-based methods for measuring poverty. The underlying notion behind this thinking was that the poor lack assets, capabilities, and initiative to take advantage of opportunities, and need to be supported with sectoral interventions. These interventions were planned as top-down programmes that looked at households as beneficiaries who lacked self-initiative.

S. Tambe, *Teaching and Learning Rural Livelihoods*, Sustainable Development Goals Series,
https://doi.org/10.1007/978-3-030-90491-3_13

The 1980s heralded the growth of participatory appraisal methods and rural livelihood approaches that looked at households located in a particular context and as proactive agents of change. These approaches took centre-stage, proliferated, and guided development policy and projects for the next few decades. This period also highlighted the role of micro-macro linkages of how along with agency of the poor, macro aspects such as structural, institutional, power, and politics play an important role in access to assets, opportunities, dignity, and empowerment. Availability of longitudinal socio-economic data enabled the assessment of the dynamic nature of household poverty. Several rural livelihood approaches such as sustainable livelihoods, governing the commons, livelihood trajectories, and graduation approach evolved and were scaled up in several parts of the globe. Conceptual advancements during this period illuminated new ideas such as the agency of the poor, livelihood capitals, common pool resources, livelihood trajectories, graduating the chronic poor, multidimensional nature of poverty, livelihood resilience, and the like. This neo-liberal development agenda was actively backed by international funding agencies and resulted in structural reforms ushering in liberalization, privatization, and globalization in the global south.

The trickle-down theory coupled with large-scale redistribution programmes not only resulted in rapid poverty reduction during the MDG period but also saw growing inequality in the middle-income countries with poverty persisting in south Asia and SSA. Both the government and market-led approaches could not address the specific issues of persistent poverty and rising inequalities. This led to the growth of rights-based approach to development which has the promotion and protection of human rights as its central tenet. These include economic, social, cultural as well as civil and political rights (Bohle, 2009). This approach is distinctive in the sense that it grounds development in the human rights legislation. It stipulates an internationally agreed set of norms that are recognized by the global community, backed by international law, and lends itself to the promise of providing a stronger basis for citizens for holding their states accountable for their duties and thereby enhancing their access to the realization of their rights (Nyamu-Musembi & Cornwall, 2004).

Another learning that emerged from practice was that those deepest in poverty were being left behind, and the imperative to look at the ultra-poor as a separate category, distinct from the poor and needing tailored interventions to graduate out of poverty. The past decade is seeing a re-emergence of macro approaches with a renewed interest in structural reforms. Globally livelihoods are in flux, impacted by rapid changes in our economic, social, and environmental systems. Over the last decade, growing vulnerabilities and crisis contexts, and their adverse impacts on livelihoods have led to conceptual advancements in our understanding of vulnerability and resilience and embedding them in development. As humanity confronts daunting environmental and health challenges in the Anthropocene, which threaten to undermine past achievements, embedding resilience in development has now become imperative to secure the future of human wellbeing.

13.2 Evolution and Sustenance of Rural Areas

It is estimated that in the year 1800, about 97% of the world's population lived in rural areas (Raven et al., 2011). The forces of industrialization, globalization, and urbanization have since then acted as a magnet attracting young adults from rural areas and 'hollowing out' the countryside (Li et al., 2019). Over the last few decades, this trend has been most noticeable in the developing world with the rural population declining by 60% in Brazil, 46% in China, 33% in South Africa, and 13% in India (Fig. 13.1). Urbanization opportunities spilled over to rural areas providing opportunities for rural non-farm (RNF) diversification. Migration and remittance got accelerated with better communication facilities, as rural livelihoods increasingly become multi-occupational and multi-locational. During the industrial revolution, there was a high dependence on the rural areas for surplus labour, capital,

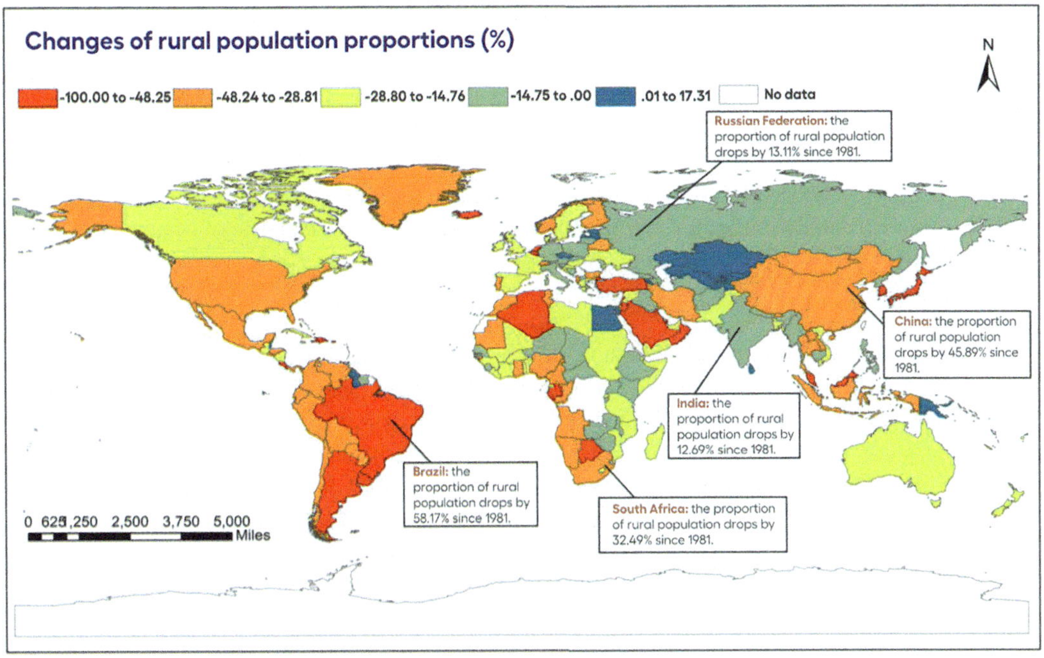

Fig. 13.1 The decline in the rural population proportion in the world (1981–2016). (Source: The World Bank data, https://data.worldbank.org/)

and resources such as land, forests, and minerals (Li et al., 2019). But in today's global knowledge economy, these resources from the rural areas have lost their relative importance.

So what are the resources that enable a village to maintain a growth pattern and not spiral into a vortex of rural decline and senescence? Li et al. (2019) suggest three necessary conditions for sustainable rural development in the knowledge economy, namely developing new economic activities that can respond to urban demand, local entrepreneurship that can establish and expand these new activities, and social capital that can enable finance, manpower, knowledge, information, and innovation. They go on to suggest that compared to the material resources of a rural community such as infrastructure and resource endowment, it is the social capital that is the key driver. As it manifests itself in the form of collective norms, trust, networks of affiliation that reduce transaction costs, enhance access to information, and resources and promote collective action. The bonding social capital involves close in-group solidarity; the bridging social capital connects diverse groups within a community; while the linking social capital connects these groups with those outside the community. Communities with a higher density of combined bonding, bridging, and linking social capital are more inclusive and participatory, more resilient in responding to changes in the external environment, and predicted to have superior development outcomes and a higher quality of life (Besser, 2009). The presence of cohesive institutions at multiple scales with strong internal and external inter-linkages enables access to opportunities, linking demand with supply, policy advocacy, and effective programme implementation.

But can programme interventions aid in building social capital? Programmes that embed institution building through interventions such as social mobilization, building local leadership, organizing women into collectives, federating them at higher spatial scales, building their social cohesion, networking them with other institutions, developing capacities, and providing access to finance to diversify livelihoods have been successful in several parts of the developing world. These programmes are at times on-scale funded by governments such as the National Rural Livelihoods Mission (NRLM) of India or at times supported by donor agencies, NGOs, and even development banks (Box 13.1).

Box 13.1 The Bonding, Bridging, and Linking Elements of Social Capital
The National Rural Livelihood Mission (NRLM) in the central Indian state of Madhya Pradesh started by investing in human capital development by organizing poor women into self-help groups (SHGs). The SHGs were formed based on the commonality of occupation, neighbourhood proximity, and the same standard of living. The capacity of these SHGs was enhanced by taking up intensive social, institutional, technical, and financial trainings, and exposure visits. They were also oriented on how to improve their health and sanitation status. At the nascent stage of SHG formation, the women members were trained on the '*panchsutra*' or the five cardinal principles of SHGs, namely regular meetings, regular savings, regular inter-loaning, timely repayment, and up-to-date books of accounts. In their regular weekly meetings, the members wore a similar coloured *sari* and also organized an opening prayer. The sense of identity was built through a common reference of the members as '*didi*' (sister). This reference has become synonymous with the members of SHG, not only with other local community members but even with government officials.

Developing this common identity helped in building group solidarity for collective action and in developing the bonding element of social capital. The bridging element of social capital is the sharing and exchange between the groups which was provided by federating the SHGs at the village level into a Village Organization (VO). The linking element of the social capital is the relationship with the higher authorities, such as government departments, which was enabled with the association with NRLM as it facilitated access to various government schemes. An independent mid-term evaluation found that the SHG members reported higher social mobility within and outside the village, could self-operate their bank accounts, had higher say in decision making, greater confidence in interaction with bank officials, and felt they were treated more respectfully. The SHG membership provided them with a new identity, which helped diminish their other identities related to caste, class, religion, and gender. Hence, the social capital of the SHG groups got enhanced due to the three elements of social capital, namely bonding, bridging, and linking addressed strongly and effectively.

Source: Patnaik et al. (2017)

13.3 Emerging Issues

13.3.1 Ending Poverty in Asia and SSA

The last two decades have witnessed the fastest reduction in poverty globally. Poverty is now concentrated in middle-income countries and is more of an inclusive growth issue involving equitable distribution rather than deprivation. Most of the poor now live in middle-income countries where the prosperity from rapid economic growth has not been shared fairly (Fig. 13.2). Poverty in these countries is more of a distribution or equity issue. Developing countries need to improve on non-income parameters of multidimensional poverty such as living standards, education, and health. The poorest now reside in SSA, which remains the final frontier in the global effort to eradicate poverty and concentrated efforts will be needed to lift them out of poverty (Fig. 13.3). There is a strong positive correlation between employment in agriculture and multidimensional poverty particularly in SSA (OPHI,

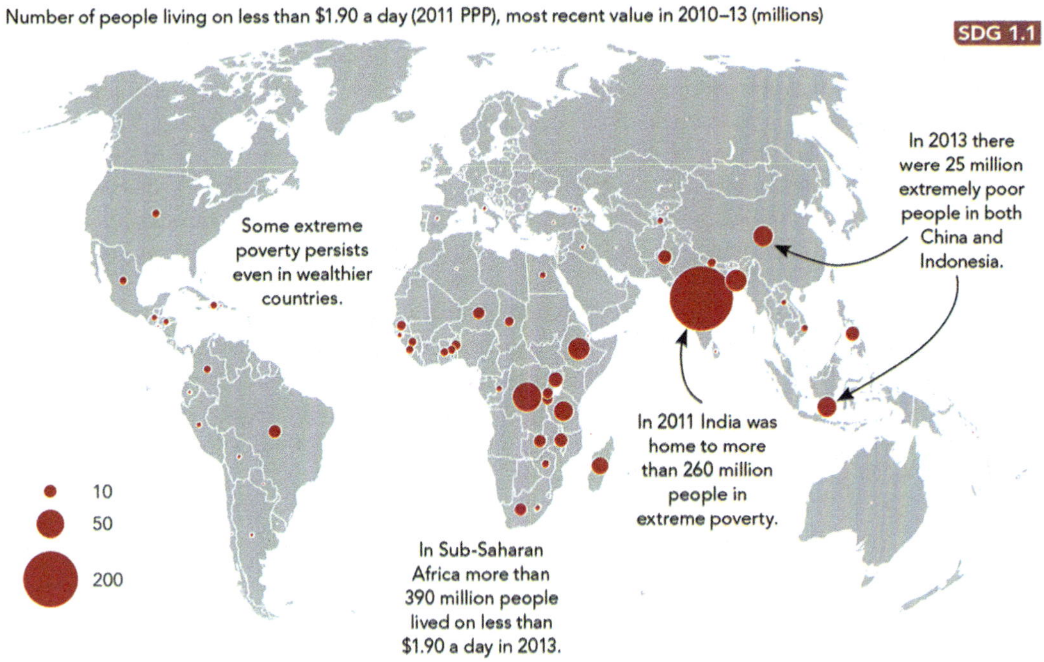

Fig. 13.2 Maximum number of poor now reside in middle income countries in the South Asia Region. (Source: The World Bank data, https://data.worldbank.org/)

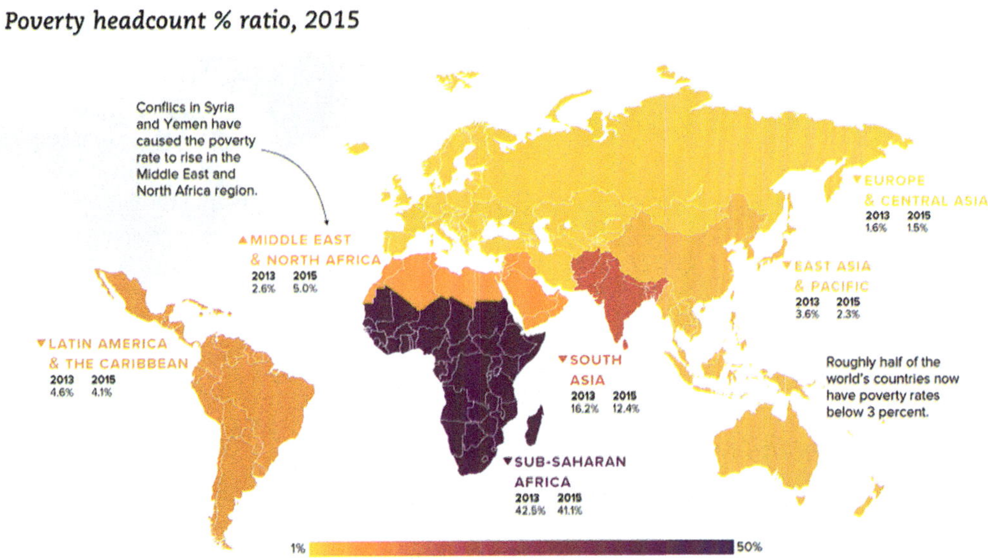

Fig. 13.3 Poverty headcount ratio (%) in 2015, showing high poverty rates persisting in Sub Saharan Africa (SSA). (Source: The World Bank data, https://data.worldbank.org/)

2020). There is also a similar positive association between a higher share in informal employment in the non-agriculture sector and multidimensional poverty. Hence, agriculture employment or employment in the RNF sector may not reduce poverty in these countries without other interventions (OPHI, 2020).

13.3.2 Creating Decent Jobs for the Less Educated

One of the biggest challenges faced by the late developing middle- and low-income countries is to create millions of decent jobs to diversify the agriculture-based livelihoods of the small holder poor. These poor have limited literacy, numeracy, and skills and are caught in the persistence of small holder farming and informal jobs in the rural non-farm sector. The creation of decent jobs has become the central issue for the political leadership in these countries where the policy priorities have evolved from survival needs (food, water, shelter, housing), to provisioning basic infrastructure (roads, electricity, water) to now decent jobs. These political priorities now figure prominently in election manifestos and the enhanced investments in skilling programmes. The main issue that policymakers are grappling with is how to unshackle the stunted structural transformation. How to create a pull factor for the less literate in urban areas by provisioning decent jobs in the non-farm sector is the challenge at hand? While provisioning of survival needs and establishment of basic infrastructure being department-driven are easier to achieve, job creation is more complex as it involves land, capital, skilled labour, and integration with the market.

13.3.3 Enabling Macro-Micro Linkages

The macro-micro link is the challenging part of livelihood development. The need to integrate the macro that comprises of transforming structures, processes, and institutions with the micro with its vulnerability context, asset pentagon, and livelihood strategy (Bohle, 2009). What is the relationship between structure and agency? Social relations of access and control and the power relations that govern the control between structure and agency need to be factored into our understanding of rural livelihoods (Bohle, 2009). Livelihood studies have become too context specific, micro, and focussed at the household level. Hence, they are unable to effectively impact sectoral social policies. Scoones (2009) highlights the lacunae in discussions regarding rural livelihoods, including questions of knowledge, politics, scale, and dynamics. Recommendations that are too context specific or have a narrow geographical spread, fail to influence policymakers. Embedding the rights-based approach in livelihood programmes can lead to the inclusion of power analysis and the potential for macro-level policy changes. He suggests bringing the politics back in livelihood approaches, by addressing macro issues both in scale and in scope to be able to better inform policymaking.

13.4 The Way Forward

13.4.1 Setting the Livelihoods Agenda

There is a need to set the agenda for the future of livelihoods research, policy, and practice. Livelihoods studies need to also diagnose the macro picture, to be able to inform social protection and structural reform policies (de Haan, 2016). The use of livelihoods in new contexts such as urban settings needs to be strengthened, especially since urbanization is an important trend in the global South. Livelihoods need to span across a range of topics such as environmental sustainability, climate change, governance, conflict, gender, migration, social protection, labour, resilience, security, and others (Mönks et al., 2017). Tackling these areas could potentially bring livelihoods up-to-date and relevant in this period of rapid global change. Also, crosscutting use of livelihood concept can be attempted in different contexts such as climate change, growing urbanization, global environmental change, water security, watershed management, sustainable development, etc. The role of markets, institutions, and technology can be built into traditional livelihood conceptualizations to improve their effectiveness in tackling poverty (Dorward et al., 2003).

13.4.2 Adding Power and Politics to Livelihoods and Strengthening Delivery

There is a need to look at the access to livelihood opportunities, power, decision making, and livelihood trajectories (de Haan & Zoomers, 2005). The new frontiers of livelihood research need to understand the relationship between structure and agency, power relations that determine access and control, and the psychological dimensions of livelihood security (Bohle, 2009). Livelihood studies need to be more concerned with informing policy and influencing action. There is a need to integrate sustainable livelihoods with the rights-based approach by putting people at the centre and taking equity, human rights, capabilities, and sustainability as its normative basis. Sustainable livelihood security views poor people as possessing a lot of agency and evaluates the dynamic portfolios of assets and the multifarious range of livelihood activities that vulnerable groups pursue to secure their lives (Bohle, 2009).

13.4.3 Alternatives to Project-Based Approach

There is a growing disillusionment with projects as agents of meeting the development needs of the poor on counts of both effectiveness and efficiency (Franks et al., 2004). This study highlights that while there have been some successes in the project-based approach, issues related to sustainability of outcomes, negative impacts, ownership by beneficiaries, high transaction costs per unit development assistance, power imbalances, and donor competition have impacted delivery. Two approaches are gaining popularity, namely Sector-Wide Approaches (SWAPs) and Direct Budgetary Support (DBS). SWAPs involve the pooling of policy advise and investment by donors across a whole sector, while DBS involves the large-scale transfer of funds directly to the treasury and it becomes part of the government budget (Franks et al., 2004). These new approaches can result in fitting the intervention closer to existing capacity, lower transaction costs, greater ownership of recipient governments, and more sustainable programme outcomes (Franks et al., 2004).

13.4.4 Strengthening the Density of Social Capital

There is a need to shift from the structural perspectives of neo-Marxism to actor-oriented perspectives with a renewed focus on the agency of the poor (Bohle, 2009). The agency of the poor can be strengthened by enhancing their bonding, bridging, and linking social capital. Building institutions of the poor such as women collectives and their federations, farmer producer organizations, milk producer cooperatives, community forest user groups, and the like can help strengthen their access and power. Livelihood development programmes need to actively engage in organizing the poor and linking them with institutions to strengthen their claim and access to their entitlements. Communities with a higher density of social capital have shown to better adapt to new threats and opportunities thereby improving human wellbeing. Projects need to embed best practices such as decentralized governance, bottom-up planning, participatory execution, and social audits to make the programmes accountable to the people (Tambe et al., 2016a). Corruption is known to impact targeting performance, asset quality, employment generation, cost effectiveness, and timeliness, thereby reducing the programme impact. Curtailing discretion, enhancing accountability, broad-basing decision making, proactive disclosure, and effective social audits can play a pivotal role in reducing corruption in programme delivery (Tambe et al., 2016b).

13.4.5 Future of Development Studies

Development studies as a discipline is in senescence or is it thriving? The continued relevance of development studies is growing, with more student cohorts expected to grow strongly in the coming years (Mönks et al., 2017). The authors highlight that the developing countries have started to question and criticize the geography of knowledge production and many concepts of modernity originating in the global north. Over the last few decades, this discipline has evolved from an 'asymmetric business', and has expanded to several developing countries in the global south (Opschoor et al., 2005). There is a felt need to embed third world experiences in the debate like China's development trajectory as a 'successful developing country' as a lesson for others as a 'south-south cooperation'. Many new ideas like microfinance, graduation approach, and others have originated from the global south. The future of development studies needs to encompass the following features—interdisciplinary nature, concern for social issues, analyse complex matters in multi-scale with multi-stakeholders and yet remain context specific, look not only at the policy but also at the processes and practice, and must be problem oriented (Mönks et al., 2017).

13.4.6 Building Resilience and Ending Poverty in the Anthropocene

Scientists are increasingly of the opinion that we have entered into a new geological epoch: The Anthropocene or the age of humans. An age where human activities have expanded to a scale so as to disrupt the very planetary processes on which their survival depends (UNDP HDR, 2020). We are the first to live in an age defined by humans, where the dominant risks to our survival are the very choices we make. In this age of 'ecological deficit' natural capital is being degraded, destroyed, and depleted at a rate faster than nature can replenish it. Not only forests, but agricultural lands, the atmosphere, rivers, and oceans are all being eroded and key ecosystem services are being compromised (Lukey et al., 2017). This ecological deficit will not only trigger mass species extinctions and destruction of wild habitats but may eventually undermine the constitutional right of the future generations to the 'Right to Life' which includes the right to a healthy and wholesome environment. This is the era that will test our determination to end poverty, our ability to embed resilience in livelihood outcomes, and thereby attain the SDG 1 of ending poverty in all forms everywhere. In this context, future livelihood development will need to track not only the societal impact but also the externality on the environmental (Fig. 13.4). The rising intensity and frequency of natural and man-made disasters in the future will test the sustainability of the SDG attainments, and the narrative to make livelihoods resilient will gain further prominence in the decades to come. The imperative for us is to ensure that the millions who have been lifted out of poverty during the MDG and SDG period, sustain their trajectory in the face of rising risks and vulnerabilities in the Anthropocene.

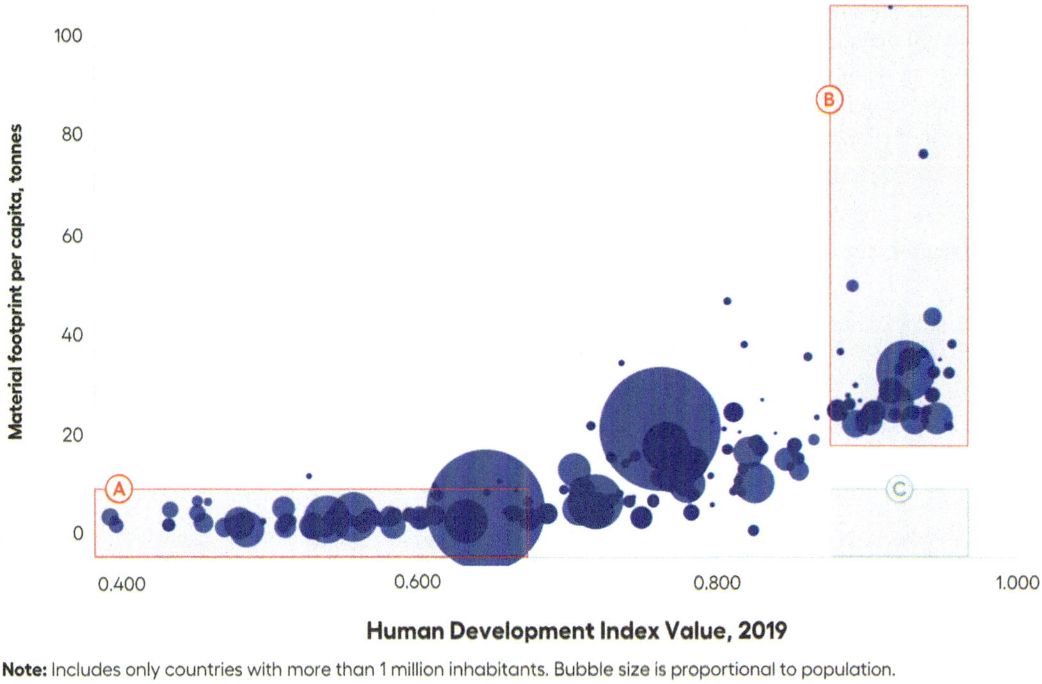

Fig. 13.4 The human development trajectory: high human development goes with high resource use. The journey of low human countries (in rectangle A) cannot follow the same high resource use pathway of the high human development countries (in rectangle B), who also cannot remain where they are. A reimagined human development pathway is now an imperative where all countries improve wellbeing equitably while easing pressures on the planet (moving to the empty rectangle C). (Source: UNDP HDR, 2020)

13.5 Concluding Remarks

The livelihood approaches need to align more closely with the strategy for achieving and sustaining the SDGs goals of ending poverty and hunger by 2030 and beyond. We opine that in the Anthropocene, the themes of livelihood trajectories, livelihood capitals, right to livelihoods, livelihood resilience, governing the commons, and graduation of the ultra-poor are more relevant than ever before. However, there is a need to adapt to new settings and contexts and also on scale to inform policy and practice. The role of politics and power in the livelihood decisions of the poor needs to be acknowledged and addressed. Social relations, property ownership, and decision making play a key role in determining access to and control over vital livelihoods resources (de Haan & Zoomers, 2005). There is a need to acknowledge that the world in which the poor live is a political arena with contested spaces, identities, and interests (Bohle, 2009). Building institutions of the poor, allowing them to shape the programmes meant for them, and crafting deliberative spaces for decision making will enable the poor to have more control over their destiny.

Typically, in academics and in practice when we think of livelihood approaches or frameworks, the DFID SLA comes to mind. In this book, we describe a family of livelihood approaches, namely the sustainable livelihoods approach, commons and livelihoods, livelihood trajectories, rights-based approach, graduation approach, and the resilience framework, and how each of them can add a unique dimension to our understanding of rural livelihoods. Based on the context, we suggest that the most suitable approach or a combination of these approaches can be adopted so that they complement each other. We opine that this strategy of picking the most suitable approach or approaches from the stable of livelihood approaches has the potential to contribute substantially to the science, policy, and practice of livelihood development. These diverse livelihood approaches can be either applied directly or embedded in standard project management frameworks. We opine that the 'family of livelihood approaches' thinking can contribute significantly in completing the task of ending poverty, sustaining the gains in the face of rising risks while simultaneously restoring the planet.

References

Besser, T. (2009). Changes in small town social capital and civic engagement. *Journal of Rural Studies, 25*, 185–193.

Bohle, H. G. (2009). Sustainable livelihood security. Evolution and application. In H. G. Brauch et al. (Eds.), *Facing global environmental change. Hexagon series on human and environmental security and peace* (Vol. 4). Springer. https://doi.org/10.1007/978-3-540-68488-6_36

de Haan, L. (2016). Livelihoods in development. *Canadian Journal of Development Studies/Revue canadienne d'études du développement, 38*(1), 22–38.

de Haan, L., & Zoomers, A. (2005). Exploring the frontier of livelihoods research. *Development and Change, 36*(1), 27–47.

Dorward, A., Poole, N., Morrison, J., Kydd, J., & Urey, I. (2003). Markets, institutions and technology: Missing links in livelihoods analysis. *Development policy review, 21*(3), 319–332.

Franks, T., Toner, A., Goldman, I., Howlett, D., Kamuzora, F., Muhumuza, F., & Tamasane, T. (2004). *Goodbye to projects? The institutional impact of sustainable livelihoods approaches on development interventions*. Bradford Centre for International Development.

Li, Y., Westlund, H., & Liu, Y. (2019). Why some rural areas decline while some others not: An overview of rural evolution in the world. *Journal of Rural Studies, 68*, 135–143.

Lukey, P., Cumming, T., Paras, S., Kubiszewski, I., & Lloyd, S. (2017). Making biodiversity offsets work in South Africa—A governance perspective. *Ecosystem Services, 27*, 281–290.

Mönks, J., Carbonnier, G., Mellet, A., & De Haan, L. (2017). *Towards a renewed vision of development studies*. International Development Policyl Revue internationale de politique de développement, 8(8.1).

Nyamu-Musembi, C., & Cornwall, A. (2004). *What is the 'rights-based approach' about? Perspectives from international development agencies*. IDS Working Paper 234. IDS, Sussex.

OPHI. (2020). *Global Multidimensional Poverty Index 2020: Charting pathways out of multidimensional poverty: Achieving the SDGs*. Oxford Poverty and Human Development Initiative. Retrieved from http://hdr.undp.org/sites/default/files/2020_mpi_report_en.pdf.

Opschoor, H., Forster, J., Jolly, R., & Mönks, J. (2005). *Development studies, accreditation and EADI*. Vision Paper presented at the EADI General Conference (Bonn 21–23 September 2005). Retrieved May 10, 2021 at 9:06, from https://www.eadi.org/typo3/fileadmin/Accreditation/Documents/EADI_Vision_paper_final.pdf.

Patnaik, S. Tambe, S., Bisaria, J., Pandey, A., Sarma, U. K., & Ananda Vadivelu G. (2017). *Evaluation of Madhya Pradesh Deendayal Antodaya Yojana – State Rural Livelihoods Mission (MPDAY-SRLM)*. Study Report, Indian Institute of Forest Management (IIFM), Bhopal, 2017, pp 205. ISBN No. 978-81-943801-1-5,

Raven, P. H., Hassenzahl, D. M., & Berg, L. L. (2011). *Environment* (8th ed.). Wiley.

Scoones, I. (2009). Livelihoods perspectives and rural development. *The Journal of Peasant Studies, 36*(1), 171–196.

Tambe, S., Subba, A. B., Basi, J., Pradhan, S., & Rai, B. B. (2016a). Measuring the effectiveness of social audits: Experiences from Sikkim, India. *Development in Practice, 26*(2), 184–192. https://doi.org/10.1080/09614524.2016.1136268

Tambe, S., Subba, A. B., Basi, J., Pradhan, S., & Rai, B. B. (2016b). Decentralizing accountability—Anti-corruption experiment from Sikkim. *Economic & Political Weekly, 51*(52), 95.

UNDP HDR. (2020). *The next frontier, human development and the Anthropocene*. Human Development Report 2020, UNDP, New York.

Index

S. Tambe, *Teaching and Learning Rural Livelihoods*, Sustainable Development Goals Series,
https://doi.org/10.1007/978-3-030-90491-3

The manufacturer's authorised representative in the EU is Springer Nature Customer Service Centre GmbH, Europaplatz 3, 69115 Heidelberg, Germany. If you have any concerns regarding our products, please contact ProductSafety@springernature.com

Printed and bound by CPI Group (UK) Ltd, Croydon, CR0 4YY
30/09/2026
02225471-0020